Anonymus

Southern Plants for Southern Homes

Evergreen Lodge materials

Anonymus

Southern Plants for Southern Homes
Evergreen Lodge materials

ISBN/EAN: 9783742804983

Manufactured in Europe, USA, Canada, Australia, Japa

Cover: Foto ©berggeist007 / pixelio.de

Manufactured and distributed by brebook publishing software
(www.brebook.com)

Anonymus

Southern Plants for Southern Homes

INTRODUCTORY.

Spring, 1893.

The cricket's croon is the last farewell to Summer, while the trill of the toad is the joyous prelude of Spring. In a little time the orchestration of small frogs, which is emphatically a vernal tone, will be heard in the land, soon to be followed by the huge green bull frog of the swamps with his grand bassoon. His nocturne brings the plash of water and the perfume of Summer flowers quite close to us. And while we listen for the tremolos of this great batrachian that so tensely emphasizes the serenity of a Summer night, we must prepare for this glad season of sunshine and flowers. The devotion to Flora as a Queen among us is now a living truth; and it is said, that a woven thread of twisted gold unites all hearts that worship at her shrine. To these then intimate friends and patrons, in anticipation of your recurring wants, we present you with this modest little Catalogue for the Spring, 1893, and trust it may be of some interest to you.

While our Catalogue does not abound in highly colored illustrations, it will be found replete with a list of plants and flowers sufficient to make all homes beautiful, and sold at prices that all may buy. Experience has now fully taught us the complete requirements of Southern flower buyers, and our facilities to meet this demand is now most perfect.

Our stock of Roses is all the most critical could desire. All varieties of recent introduction and established merit are offered at popular prices. Our stock of Chrysanthemums has no equal in this country. We have revised our list of varieties, discarding hundreds of older sorts now superseded by varieties of more recent introduction, so that our list of Chrysanthemums to-day does not contain a single variety but what is of high merit, while on the other hand there is not a variety known to plant commerce that is worth growing but what will be found enumerated in our Catalogue.

We here wish to thank our patrons for past favors, and respectfully solicit a continuance of the same, and trust by honest treatment and courteous attention to continue this connection to the mutual welfare of both. With best wishes of a happy New Year to all, we remain, most truly,   JAS. MORTON, Manager.

J. J. CRUSMAN, Proprietor.

A Few Points—Read Before Ordering.

Our Terms.

Our terms are invariably Cash with the order.

Visitors Welcome.

Visitors are always welcome, as we have something of interest for them to see at all seasons of the year.

Express.

All orders for goods not stating the mode of transportation will be sent by Express at the purchaser's risk.

Remittances.

In remitting, send Money Order or Draft; if in currency, invariably register the letter, as we will not be responsible for remittances otherwise made.

Postage Stamps.

When you cannot procure a money order and cannot make change otherwise, we will accept Postage Stamps. We would prefer two, five and ten cent stamps.

Postal Notes.

Postal Notes are very convenient for small amounts, but are no more safe than money sent in an ordinary letter, as they cannot be duplicated if lost, and anybody that gets them can collect them.

Our Shipping Facilities.

Our shipping facilities are first-class, and the rates by express or freight from this point are much less to all the States South of us than they are from any of the Northern floral establishments; the time occupied in transit is also much less.

Have Your Goods Sent by Express.

If possible, have your goods sent by express, is always our advice, as you invariably get your plants in a better condition; and, as an inducement, we send a lot of plants extra to help pay express charges, and we feel ourselves repaid by the better satisfaction our shipments will give you.

Orders For Less Than One Dollar.

No orders for less than one dollar will be sent by mail unless fifteen cents additional to the price of the plants be sent to pay postage. It is quite as much trouble to handle, and requires as much postage to mail, a fifty cent order as it does one for two or three dollars' worth of plants.

When Our Liability for Loss Ceases.

We take a receipt for all plants delivered in good order to carrying company, when our liability ceases, and the plants are at the risk of parties ordering. We make no charge for drayage or packing. If mistakes occur, notify us at once; otcerwise, we are not responsible.

Letters and Packages.

Letters travel somewhat faster in mails than packages, so, if we write you a letter, and it reaches you before the plants, wait a day or two before writing, and give them the necessary time, and in ninety-nine cases in every hundred all will come out right, saving both of us the trouble of writing.

Be Patient.

In our busy season the office work is so pressing that packages of plants frequently leave the greenhouse before we get an opportunity to write, and as this is unavoidable, we beg our customers, if any plants are missing, to kindly wait two or three days for a letter of explanation before informing us of the shortage.

C. O. D. Orders.

C. O. D. orders must be accompanied by at least one-fourth of the amount in cash, and the parties ordering are to pay the express charges for collecting. Large orders for shrubbery, trees, etc., can go by ordinary freight, by consigning to our own order and sending bill of lading by express, C. O. D., endorsed to the parties ordering. Heavy express charges are thus avoided and collections facilitated.

Wholesale Trade.

We have constantly a demand from florists and dealers in the South for our wholesale trade list, and here wish to state we issue no Wholesale Trade List. We grow our plants in large quantities, and are able to give the lowest trade price on all goods we offer to florists and dealers. To all such purchasers, we say send us a list of what you want, and we will quote special wholesale rates, and can do as well for you as the so-called wholesale florists that sells to the florists and the florists' customers at the same price.

Mistakes.

If anything is wrong with your order, do not think we intended it, for we have no interest in so doing; our interest is to give satisfaction, and that we are determined to do; so should an error occur, kindly put it down as a mistake, notify us, and we will put it right. We wish our customers would, in every case, keep a copy of their order, and verify it on arrival of plants; this will prevent mistakes as to what they "thought" they ordered, but which was never written upon their order sheet; and if not too much trouble, please drop us a card on the arrival of the goods. We are glad to know when you are pleased, and we wish to know of any dissatisfaction, that we make it right.

Addresses.

Please be careful and write your Name and Address plainly. We can readily make out what is wanted or what we are acquainted with the names of all our plants, but we have no means of knowing what your name is or how it should be spelled unless you write it plainly. We receive many orders that are well written throughout, but when we look to see who sends them, the name is so carelessly written that we are obliged to guess at it. Indeed, some forget entirely to sign their name. Again we would say, "please sign your name carefully," as it will save us much annoyance, and possibly, prevent errors.

Inducements to Clubs.

ALTHOUGH prices are low in this Catalogue for all classes of plants, most liberal terms are offered to friends who are inclined to obtain orders from others, and to secure thereby some fine specimens free of cost for themselves. In making up a club order it is important to state the sum sent by each member and the plants wanted, that they may be separately packed and confusion avoided when the plants are distributed. The full address of each is always required. The following are the rates from $35.00 to $2.00 (larger sums in proportion):

For a $3.00 club order the originator may select in plants 50 cents.
For a $4.00 club order the originator may select in plants 75 cents.
For a $5.00 club order the originator may select in plants $1.00.
For a $6.00 club order the originator may select in plants $1.10.
For a $7.00 club order the originator may select in plants $1.25.
For a $8.00 club order the originator may select in plants $1.75.
For a $9.00 club order the originator may select in plants $1.75.
For a $10.00 club order the originator may select in plants $2.00.
For a $15.00 club order the originator may select in plants $3.00.
For a $20.00 club order the originator may select in plants $4.00.
For a $25.00 club order the originator may select in plants $5.00

Cheap List.

THE following collections, to be sent by express only, are very desirable to those who want a nice flower bed and care nothing about having the names put on each plant, the doing of which during the busy season consumes valuable time. We desire it distinctly understood that the plants in these collections are just as good and desirable in every way, and probably would be better than the individually selected plants at more than double the price.

One Dollar Collections.

Owing to a large increase in our facilities for raising plants this last Summer, we are able to offer plants in the following collections at the exceedingly low rate of twenty-five plants for one dollar, by express only, and no premiums with these collections. If wanted by mail, add fifteen cents extra:

25 Hollyhocks.	25 Violets.	25 Nasturtiums.
25 Verbenas.	25 Heliotropes.	25 Ageratums.
25 Coleus.	25 Tuberoses.	25 Salvias.
25 Pansies.	25 Gladiolus.	25 Cannas.
25 Geraniums.	25 Chrysanthemums.	25 Tulips.
25 Achyranthus.	25 Centaurea.	25 Petunias.
25 Carnations.	25 Asters.	

If the parties ordering prefer, they may select five plants from five of the different collections; thus, five Pansies, five Geraniums, five Tuberoses, five Chrysanthemums and five Verbenas. Not less than five plants from any one collection to make up the twenty-five as offered for one dollar. The plants will be packed nicely in a small basket, and sent by express. Six collections for five dollars.

Collection of Plants for Five Dollars.

By Express Only.

The following collection, containing one hundred and fifty-six plants, for five dollars, is the cheapest ever offered, and every one of them are fine strong plants that will grow rapidly and make beautiful any home and its surroundings for the Summer, and many of them will stand the Winter and come again the following season. Nobody should neglect to beautify the surroundings of their home with an offer like this before them. The selection of all the varieties must be left with us. We cannot hunt up named varieties at this price. They will all be packed carefully in a light box or basket, and sent anywhere by express for five dollars. One-half the collection for three dollars. No premium with this collection:

6 Everblooming Roses.	6 Pansies.	4 Violets.
6 Chrysanthemums.	6 Hyacinths.	4 Rose Geraniums.
6 Achyranthus.	6 Paris Daisys.	4 Lantanas.
6 Fuchsias.	6 Verbenas.	4 Abutilons.
6 Geraniums.	6 Asters.	4 Sweet Allysum.
6 Tuberoses.	6 Ageratums.	2 Evening Glories.
6 Carnations.	6 Coleus.	2 Plumbagos.
6 Heliotropes.	4 Jasmine.	2 Hibiscus.
6 Gladiolus.	4 Tradescantias.	2 Dahlias.
6 Begonias.	4 Hollyhocks.	2 Callas.
6 Salvias.	4 Feverfews.	2 Lantanas.

One Dollar Mail Collections.

Look carefully at this offer for one dollar. There are many bright and happy homes throughout the South where intelligence is supreme, and consequently good flowers appreciated, that are not fortunate to have an express office convenient to them. To place our flowers within the reach of such people, we have prepared the following collections that we will send free, postpaid, through the mail for one dollar, packed carefully in a nice wooden box. Any one of these collections will make a handsome bed, and nothing helps to make a home more cheerful than a neat flower garden, however small. If preferred, parties may select four plants from any five collections, and make up their twenty plants in that way. Any three collections for $2.50, or six for $5.00:

20 Geraniums.	20 Gladiolus.	20 Heliotropes.
20 Chrysanthemums.	20 Verbenas.	20 Coleus.
20 Carnations.	20 Asters.	20 Achyranthus.
20 Tuberoses.	20 Fuchsias.	20 Violets.
20 Pansies.	20 Salvias.	

Mixed Mail Collections for One Dollar.

Please Order by Number.

1 —4 Geraniums, 4 Roses 3 Coleus, 3 Heliotropes, 3 Pansies, 3 Verbenas.

2 —4 Fuchsias, 4 Carnations, 2 Plumbagos, 4 Salvias, 3 Tuberoses, 3 Gladiolus.

3 — 6 Violets, 6 Daisys, 4 Pansies, 2 Hollyhocks, 2 Abutilons.

4 —4 Lantanas, 4 Petunias, 2 Crape Myrtle, 3 Begonias, 4 Smilax, 3 Chrysanthemums.

5 —4 Roses, 4 Carnations, 4 Hollyhocks, 3 Scented Geraniums, 3 Tuberoses.

6 —6 Carnations, 6 Pansies, 6 Asters, 1 Calla, 1 Lily of the Valley.

7 —6 Chrysanthemums, 2 Violets, 2 Roses, 2 Abutilons, 2 Salvais, 6 Fuchsias.

8 —2 Crape Myrtle, 2 Feverfew, 4 Roses, 4 Hollyhocks, 2 Violets, 6 Garden Pinks.

There is not a home in the South or a person that receives this Catalogue, but what can use at least one of these collections to advantage. If you are boarding at a hotel and have no place to put them out, they will make a nice present for a friend not so situated, and will afford pleasure and remembrance all the Summer long. We ask therefore as an acknowledgment that this Catalogue is appreciated, an order for at least one dollar's worth of flowers, so that your name may go permanently on our books as customers, and continue to receive our catalogues.

ROSES.

S far back as can be remembered, the Rose has been acknowledged as the "Queen of Flowers." No garden, however small, is complete without Roses. There are no flowers grown that are more universally admired than the Rose, and their cultivation is yearly extending, as it becomes more generally known that they are so easily grown, and that they can be procured at so trifling an expense. All that is necessary is to plant them in a bed of deep, fresh, loamy soil, well enriched with thoroughly rotted manure, and they are as certain to do well as a bed of Geraniums.

CULTURAL NOTES.

PREPARATION OF THE GROUND.

Roses will grow in any fertile ground, but are much improved in bloom, fragrance and beauty by rich soil, liberal manuring, and good cultivation. The ground should be subsoiled and well spaded to the depth of a foot or more, and enriched by digging in a good coat of cow manure or any fertilizing material that may be convenient. Renew old beds by decayed sods taken from old pasture land.

PLANTING.

When the ground is thorough prepared, fine and in nice condition, put in the plant slightly deeper than it was before, spread the roots out evenly in their natural position, and cover them with fine earth, taking care to draw it closely around the stem, and pack firmly down with the hand. It is very important that the earth be tightly firmed down on the roots. Budded Roses should be planted three inches below the bud. Always choose the most favorable time for planting in your own locality. Roses can be planted as soon as convenient after the frost is over. Always select an open, sunny place, exposed to full light and air. Roses appear to the best advantage when planted in beds or masses.

WATERING.

If the ground is dry when planted, water thoroughly after planting, so as to soak the earth down below the roots, and, if hot or windy, it may be well to shade for a few days. After this not much water is required unless the weather is unusually dry. Plants will not thrive if kept too wet and without drainage.

PRUNING.

In most seasons it is best to prune established plants of hardy kinds in February. Tender varieties, such as the Tea Roses and newly planted Roses, may be left till a month later. As a general rule close pruning produces quality, and long pruning quantity of bloom. Climbing, Weeping and Pillar Roses should not be cut back; but the tips of the shoots only should be taken off, and any weak or unripe shoots cut out altogether.

INSECTS AND DISEASES.

THE APHIS.

The Aphis or Green Fly is well known to all who have grown Roses. It is a small green louse about one-eighth of an inch in length when fully grown, and through their slender beaks they suck the juices of the plant, always working at the tender shoots, and in a short time will, if unmolested, destroy the vigor or vitality of any Rose they infest. The best destructive agent to use against them is tobacco; if growing in a pit or greenhouse it may be burned so as to make a smoke. Care must be taken not to smoke it too much; better light applications and repeat a couple of times until the fly is dead. If the plants are grown out of doors, and infested with fly, a liquid solution made from tobacco stems will be found an efficient method of working their destruction. Take some tobacco stems and place in a tub or vessel of some kind, and pour boiling water upon them until the liquid has the color of strong tea; after it cools off sufficiently to handle it, apply it to the Rose with a syringe or wisk broom; a little soft soap or whale oil soap added to the solution will greatly aid it in its efficacy.

MILDEW.

This is a fungous disease often caused by great and sudden atmospheric changes and a long continuance of damp cloudy weather. The best remedy is sulphur, and should be applied the moment the disease makes its appearance, which is in the form of a white or grayish substance covering the leaves and causing them to crimple and become deformed. The plants should be sprinkled first with water so that the sulphur will stick; the best plan though is to apply it in the morning while the dew is upon the plants. After a few days the sulphur will all fall off and the mildew disappear. This treatment applies to Roses which are grown both in-doors and out, but if grown in a pit or greenhouse the best way is to mix the sulphur with water to the consistency of a good stiff paint, and apply it to the pipes or heating apparatus in the house with a brush. The fumes given off from this will at once check the ravages of the mildew.

RED SPIDER.

This is a most destructive little insect, and generally commits its ravages in a greenhouse or pit, and only make their appearance when favored by a hot and dry atmosphere. These are very small, scarcely distinguishable by the naked eye; if isolated, they are of a dark reddish brown color, found on the under side of the leaves, and cause the foliage to assume a yellow tinge, and soon make lickly the plants they infest. A few applications of whale soap dissolved in warm water, mixed with tobacco water, applied with a syringe and thrown upward so as to strike the under side of the leaves, will soon destroy them. This insect does not attack plants that are syringed with water daily, and all plants grown under glass, not in flower, should be sprinkled overhead with water daily.

BLACK SPOT.

This disease seems to be troublesome in many places, and Rose growers in the Northern States have suffered much from its ravages. It has of late made its appearance in many places in the South, although at present it is not generally known. The Hybrid Perpetuals and the Hybrid Teas appear to suffer most from it. As its name implies, it is a black spot that comes upon the leaves of the Rose, and gives it a somewhat blighted appearance. As soon as the plant becomes infested with it, it loses all its vigor and will cease to make further growth. The real cause of Black Spot is at present a disputed question. Since the causes from which it emanates is so badly understood, it is of course equally difficult to suggest a remedy. When grown in greenhouses, the best means of checking the disease we have found is a healthy, dry atmosphere at night and a free circulation of air during the day, with a little fire heat to counteract any cold draughts. Where Roses are infested with Black Spot in the open ground, the best remedy is to cut the plants back and remove all leaves infested; when it starts to grow again the chances are that the Black Spot will not appear.

ROSE HOPPER.

This is another troublesome pest with which the Rose is afflicted in the open ground. It is a small yellowish white insect about three-twentieths of an inch long, with transparent wings. Like the Red Spider they prey upon the leaves, working on the under side. They go in swarms, and are very destructive to the plant. As they jump and fly from one place to another, their destruction is less easy to accomplish than is the case with other enemies. Syringing the plants with pure water, so as to wet the under side of the leaf, and then dusting on powdered hellebore or tobacco dust, will destroy or disperse them.

ROSE SLUG.

These slugs are the larva of a saw-fly, about the size of a common house-fly, which comes out of the ground during May and June. The female flies puncture the leaves in different places, depositing their eggs in each incision made; these eggs hatch in twelve or fifteen days after they are laid. The slugs at once commence to eat the leaves, and soon make great inroads upon the foliage if not checked. They are about one-half inch long when fully grown, of a green color, and feed upon the upper portion of the foliage. The best remedies are powdered white hellebore, or a solution of whale oil soap.

ROSE CATERPILLAR.

These are the young moths or butterflies, varying from one-half to three-fourths of an inch in length. Some are green and yellow, others brown. They all envelop themselves in the leaves or burrow in the flower buds. Powdered hellebore will prevent in a large measure their moving over the plants, but the only method of killing them that is really effective is picking them off with the finger and thumb and then tramping them under the foot.

Roses from Five Inch Pots.

We grow the following list of Roses in five-inch pots. They are all large two year old plants, very bushy, and varying in height from twelve to twenty-four inches, according to the habit of growth of the different varieties. This list comprises many of the newer varieties of recent introduction and excellent merit, and are offered here at the same price as all the old standard varieties. This list comprises Teas, Hybrid Teas and Hybrid Perpetuals, and for the size of these plants, the excellent assortment of varieties offered, and the low price when both these facts are considered, we think this list has no equal in any Catalogue of the present season.

Price 25 Cents Each ; $2.50 per Dozen.

Augustine Guinoiseau. Or White La France, with which it seems to be identical in habit, form and vigor, but in color it is a delicate blush at the centre, growing lighter and lighter until along the edges it is almost a pure white.

Arch Duke Charles. Color a brilliant crimson scarlet with violet.

Agrippina. A rich, velvety crimson; beautiful buds; for bedding is unsurpassed; few Roses are so rich in color.

Beauty of Stapleford. Flowers very large and of perfect form; deliciously scented. Tea fragrance; color a clear, bright pink, shading to a bright rosy crimson; makes large and beautiful buds, and is a constant and profuse bloomer.

Bride. This is a sport from Catharine Mermet, of which it is a facsimile, except in color, which is pure white; it is a more abundant bloomer than the parent, and as a cut flower lasts longer fresh than any other white Rose in existence.

Bon Silene. Is equally valuable for Summer or Winter blooming; buds of beautiful form; an unusually free bloomer; color deep rose, shaded carmine.

Bona Wellshott. A very strong grower; flower large and double, and of the centifolia form; color rosy vermilion, with centre of orange red.

Countess Anna Thun A strong bushy grower, with flowers freely produced on short stems; flowers extra fine and large; color a rich orange yellow, shaded with silvery salmon.

Chateau des Bergeries. Large canary yellow bud, nearly equal to Perle des Jardins in size.

Climbing Perle des Jardins. In this new sort we have all the good qualities which made its parent famous, viz: strong, healthy constitution, freedom of bloom and the delightfully fragrant deep yellow flowers, familiar to every lover of choice Roses; when to these excellent traits we add that the new Climbing Perle is a most rampant running Rose, some idea of its great value may be had.

Clotilde Soupert. In this beautiful variety the plant grows from sixteen to eighteen inches high, and is an excellent sort for either bedding or pot culture, the flower is large, very full and finely imbricated, the outer petals are pearly white, shading to a fine rosy pink centre; very free flowering.

Countess de Vitzthum. Strong growing, bearing a profusion of large, double, finely formed flowers; color of centre a bright Naples yellow, shading to a lighter tint

Catharine Mermet. One of the finest Roses grown; the buds are very large and globular, the petals being recurved and showing to advantage the lovely bright pink of the centre, shading into light creamy pink, reminding one of La France in its silvery shading.

Capt. LeFort. Color of flower a rosy purple; reverse of the petals a china rose; long buds on stiff stems; a variety of great merit.

Coquette de Lyon. A good growing variety and very free flowering; a fine yellow Rose; called the yellow Hermosa from its free flowering habit.

Cornelia Cook. A beautiful creamy white; buds of immense size and very double; does not open well at all times, which is its weak point, but when well grown is a magnificent flower.

Countess Eva Starhemberg. The bud is long in form, opening into a fine double flower of very great beauty and heavy texture; color a creamy yellow, shading to ochre at the centre; borders of the petals touched with rose.

Charles de Franciosi. The bud is long and nicely formed, of orange red color; the flower is large and double; color a chrome yellow, shading to salmon, the outer petals shaded rose.

Comte de Paris. Brilliant red, shaded and illuminated with bright crimson; large, full and of fine form; very vigorous.

Comptesse de Frigneuse. A very fragrant and beautiful Tea Rose with long pointed buds like Niphetos; color delicate canary; very free bloomer and a fine forcing variety.

Dr. Dusillet. A strong dwarf grower, with large flowers of salmon yellow color at the centre, changing to clear yellow at the ends of the petals, also shaded rose; very free in bloom.

Duchess of Edinburg. A splendid Rose, in great demand for its lovely buds, and remarkable for its beautiful color, which is one of the most glowing scarlets.

Duchess de Bragance. Light canary yellow; after the style of Coquette de Lyons, but stronger and of better construction.

Dr. Grill. Medium size, vivid yellow, centre light orange, shaded pink; very exquisite fragrance.

Duchess of Albany. This variety is a sport from La France, but is far superior to it in every way, deeper in color, more expanded in form and larger in size; the flowers are deep, even pink, very large and full.

Edmond Sablayrolles. Growth vigorous, producing short stems, crowned with flowers; color of flower a beautiful hortensia rose; interior of the flower reddish peach color, shading finally into rosy carmine.

Elize Beauvillalue. A variety especially valuable for the Southern States, where it will be entirely hardy; it is a very rapid grower and produces deep buff colored flowers, very similar in color to the well known Sunset Rose.

Emile Bardiaux. A strong, rampant grower; flowers large and double; color a bright carmine red, shaded with soft violet.

Euphrouse. Pale yellow, orange tint; a free bloomer and good grower.

Eliza Fugier. A seedling from Nipbetos, which it greatly resembles in form and bud, while the color is a deep cream, sometimes edged pink; it is very free in bloom, has handsome foliage, and is of better habit than the parent.

Etoile de Lyon. This we consider one of the finest yellow bedding Roses for outside planting, and one of the hardiest in the Tea section; flower very large and double and deliciously fragrant; color chrome yellow, deepening in centre to pure golden yellow.

Esmeraldo. A very vigorous and robust growing variety, producing dense, light colored, green foliage; exceedingly free flowering; flowers medium sized and well formed; color silvery flesh and shaded with fawn; very beautiful.

Earnst Metz. Robust in growth, but dwarf in habit; flowers very large when open; long and pointed buds, produced on long stems; color a rosy carmine, with the color heightened in the centre.

Fursten Bismarck. A very variable colored Rose, changing from china rose color to cherry red, the whole suffused with lemon; strong and vigorous.

Glorie de Margottin. A dazzling red, the most brilliant yet known; large, full and finely formed globular flowers; growth very vigorous, and can be specially recommended for its vigor of growth, freedom of blooming and hardiness.

Gustave Piganeau. Flowers are extra large, equaling Paul Neyron in size, double, and of cup form; color a beautiful shade of red and brilliant carmine.

Golden Fairy. Golden color; of dwarf habit; most beautiful of all Polyanthas; flower about the size of a twenty dollar coin and borne in the greatest profusion.

Gustave Nadaud. A free branching grower, with large double flowers; the outside petals are large and rounded, giving it an exquisite cup shape; color vermilion, with clear touches of carmine lake and soft pink centre.

G. Nabonnand. A very strong, rampant grower; flowers large, nicely formed, petals unusually large; color rosy carmine, shaded silvery yellow; delicate in tint.

Gloire des Polyanthus. A beautiful dwarf variety, with quite small flowers, which are prettily cupped; a real fairy Rose; the color is a bright pink, with a red ray through each petal; quite distinct.

George Pernet. A strong growing and dwarf variety, forming a perfect round bush; it is continually in bloom, the flowers being quite large for this class; the color is bright rose with touches of yellow, and passes to peach-blow with silvery white shadings.

Hermosa. Always in bloom and always beautiful; the flower is cupped, finely formed and full; color the most pleasing shade of pink; very fragrant; a favorite with every one.

Jeanne Guillaumez. A very vigorous grower; flower large and double, of good form, with beautiful long buds; color clear red, touched with salmon; centre coppery red, with pale silvery shadings.

J. B. Varrone. A fine grower; flowers large and very double, with high centre opening from long buds; color a soft china rose, changing to a bright deep carmine of even shading; an extra good Rose and very sweet.

Joseph Metral. A very strong, healthy grower; flowers unusually large and of good form; color magenta red, passing to cerise red, shaded with purple; received two medals of merit.

Lady Castlereagh. Flowers very large, full, beautifully formed; petals thick, round and smooth; color soft rosy yellow, with rose color predominating on the outer edge of petals.

Luciole. A very bright carmine rose, tinted and shaded with saffron; base coppery; back of petals bronze; large, full and finely scented; of good shape; long buds.

Lady Arthur Hill. A most vigorous grower, with flowers of the largest size, double, and finely formed; color a fine silvery rose of most pleasing shade; a seedling from Beauty of Waltham.

La France. Splendid satin rose, very large, full and of fine form; a constant bloomer; the sweetest of all Roses, and none can surpass the delicacy of its coloring.

Little Pet. As it opens the bud appears a bluish color, but it is soon seen that it is only on the back of the outer row of petals, the other portion of the Rose being pure white.

Letty Coles. One of the loveliest Roses grown without exception; it is very beautiful and cupped, forming a magnificent full, open Rose, of soft, creamy white, with very bright, rosy carmine centre; a splendid grower, and a universal favorite.

Mme. Margottin. Dark citron yellow, carmine centre.

Mignonette. One of the most beautiful miniature Roses imaginable; the flowers are full and regular, perfectly double, borne in large clusters, often thirty to forty flowers each; color clear pink, changing to white, tinged with pale rose; a constant and profuse bloomer.

Mad. Etienne. The flowers are large and very double; the color a delicate pink, deeper on the edge of petals; very free flowering; an excellent bedding sort, blooming the whole season.

Mrs. James Wilson. Flowers large and double, deep cream color; edge of petals touched with soft blush; flowers upright on strong, stiff stems; an elegant bedding sort.

Mme. Renahy. Flowers large, double, and of fine globular form; color a rosy carmine, with brighter centre; reverse of petals soft silvery heliotrope; very sweet and free.

Marie Ducher. Rich, transparent salmon, fawn centre; large and distinct.

Marchioness of Lorne. Flowers of an exceedingly rich and fulgent rose color, slightly shaded in the centre with vivid carmine; they are large, very sweet, full and of finely cupped shape; the petals are large and the buds long and handsome.

Mad. Georges Bruant. This new ever-blooming Rose inherits the beautiful foliage and hardiness of the Rugosa, with the flowering qualities of the Tea class; color pure white; very fragrant; buds long and pointed.

Mad. Pierre Guillot. Moderately vigorous in growth; flowers large, of nice shape; color orange yellow, bordered and lined with rosy carmine; reverse of the petals yellowish white.

M'me Hon. Defresne. Beautiful dark citron yellow, with coppery reflex; as charming in bud as in open flower; a strong grower and free bloomer.

Mad. Hoste. This is one of the finest Tea Roses that we have had for years; it is a strong grower, and buds can always be cut with long stems.

M'me Schwaller. This fine Rose has the strong, firm growth of the Hybrid Perpetuals; the color is a bright rosy flesh, paler at the base of the petals.

Marie Lambert. This is called "Snowflake" by some growers. It is a vigorous grower and free bloomer; a pure white in color.

May Rivers. A white Tea; outer petals creamy white, centre clear lemon yellow; large, deep, finely formed blossoms.

M'me Scipion Cochet. In bud yellowish pink, banded flesh rose, centre yellowish; a very free bloomer.

Meteor. A Hybrid Tea, of strong bushy growth, producing quantities of finely formed deep crimson-scarlet flowers; a very free and productive variety.

M'me Francisca Kreuger. Extra fine, orange yellow, tinted rose.

Mad. de Watteville. The color is a remarkable shade of creamy yellow, very richly tinged with carmine, with large, shell-like petals.

Mme. Martha du Bourg. The outer petals recurve at the edges, showing a nicely pointed centre; the color is a creamy white, touched with carmine or pale heliotrope on edges; texture heavy.

Madeleine d'Aoust. A strong, medium sized bush; flower large and double; of extremely fine form; primrose yellow.

Mrs. Paul. This is a Bourbon of vigorous habit and produces flowers of great beauty and exceptional distinctness and substance; color a pearly white, suffused with peach.

Miss Edith Gifford. Flowers large, very fine both as bud and open flower; color creamy white, with very distinct light pink centre.

Mons. Fruitardo. Bright sulphur yellow; good, full form and fragrant.

Marie Sisley. Pale yellow, margined rose; deliciously scented.

Mad. Lambard. Extra large full flowers; very double and sweet; color a beautiful shade of rosy bronze, changing to salmon and fawn, shaded with carmine; buds and reverse of petals a deep rosy crimson.

Mad. Cusin. Purplish rose, the centre slightly tinged with yellowish white; very distinct; flowers large, full and well formed; very fine.

M. L. de Vilmorin. A strong grower; flower large and double, and of especially fine form; clear bright red, with dark veinings and shadings of velvety brown; a variety of the very highest order.

Martin Cahuzac. Flowers of extra size and very fine form, quite globular; a beautiful rose color and bright carmine; extra good.

Mme. Ph. Kuntz. A strong grower, bearing its flowers quite erect; these are large and double, of a cherry red color, passing to salmon and flesh.

Mad. Emile Vloeberghs. Of vigorous growth; fine, well formed flowers, produced on short shoots; color pale yellow, shaded with rose and vermilion.

Marie Guillot. White, tinged with a delicate shade of lemon; large, full and beautifully imbricated in form; one of the finest white Teas.

Princess Sagan. A strong growing Rose; small, closely set, dark foliage; medium sized flowers of the brightest scarlet and velvety texture.

Parquette. One of the finest of this class; the flowers are pure white, of the most perfect shape; about one and one-half inches in diameter, flowering in clusters of from five to thirty flowers each; a very free bloomer, and one of the finest pot plants.

Perle des Jardins. Unquestionably the finest yellow Rose for either Winter or Summer flowering; the flowers are very large and double, of perfect form; color a rich shade of yellow; a healthy, free grower, with beautiful foliage, and unequaled in profusion of bloom.

Papa Gontier. It is a strong grower, with fine healthy foliage; the buds are large and long, with thick, broad petals of a dark carmine crimson color, but changing to a lighter shade in the open flower; an excellent Winter blooming variety, and one of the best for outdoor planting.

Prince Bismarck. Clear, brilliant yellow, producing fine, large buds of golden yellow balls; a strong and vigorous grower; this is a seedling from Gloire de Dijon.

Princess Beatrice. A most beautiful Rose, and will be of value to all florists for Summer buds; splendid for pots and fine for bedding.

Progress. Color a superb brilliant rosy carmine, with yellow shadings at the base of the petals; the flower is large, nearly double; very free in flower and a fine grower.

Souv. de Gabrielle Drevet. Salmon white, rose centre.

Red Malmaison. Like this well known variety; of a rich deep red; a strong grower.

Rainbow. This new variety is a sport from that excellent sort Papa Gontier, and is striped and splashed with carmine on a beautiful pink ground; it is a decided and interesting novelty, everblooming like its parent.

Souv. Clairvaux. Flowers medium to large in size, and of beautiful form; color bright china rose, the base of the petals apricot, shading to nankin yellow, with touches of carmine; free in bloom and very sweet.

Souv. Victor Hugo. Bright china rose, with copper-yellow centre; outer petals suffused with carmine; a beautiful combination of coloring.

Souv. d'Auguste Legros. A very fine grower; flowers large and double, with beautiful long buds; color a fiery red, mingled with silvery crimson.

Souv. d'F. Gaulain. The color varies from magenta to silvery violet or crimson; a nice habited grower, with large double flowers.

Souv. de Wooten. An American variety of great promise which is being largely planted; color rosy crimson, or crimson red.

Sombreuil. Large, fine formed flowers white, tinged with delicate rose; fine blooms in clusters.

Susanne Blanchett. Foliage is large, erect and beautiful; outer petals large and broad; of a clear, bright pink.

The Queen. A most charming white sport from Souv. d'un Amie; will be useful for white flowers.

Triumph de Pernet Pere. Of strong constitution and fine habit; the flower large, double and of heavy texture; bud long and nicely pointed; free in bloom.

Viscountess Folkestone. A Hybrid Tea Rose of artistic shape and color; it forces readily, is of strong growth and bears lovely white flowers tinged salmon pink; is slightly cupped and delightfully fragrant.

Valle de Chamounix. The coloring of this Rose is simply exquisite; the base and back of the petals are a bright yellow, the centre highly colored with glowing copper and rose.

Waban. A Red Mermet. Rich deep bright pink; the Waban retains its deep rich color in all kinds of weather; it will, without doubt, prove to be as valuable as The Bride, which is also a sport from the same fine variety.

Novelties of the Year.

Annually the European Rose growers send out from fifty to one hundred new varieties. Only a few prove valuable and of superior merit. We buy all that are offered that are suited to the American trade. We know that the Roses offered below contain much of promise to the American grower, the Hybrid Teas being specially promising for the Southern States.

Price 50 Cents Each.

Kaizerina Augusta Victoria.

Hybrid Tea. A German variety, and one of great interest to the Rose forcer; color, pure ivory white. A large full flower, magnificent in bud. Stems long and bearing large, strong canes, with handsome foliage. It did finely for us all last Winter. We believe this Rose to be a valuable addition to our forcing varieties.

Danmark.

Hybrid Tea. A seedling from La France, introduced from Denmark. This has all the good qualities of its parent, but is different in form and shape; looks more like a finely-built Hybrid Perpetual. A strong, vigorous grower, producing flowers as freely as La France.

Augustine Halem.

Hybrid Tea. This is like Marie Baumann in form, color and shape, but produces flowers with the freedom of a Tea. Color, dark crimson. This is one of Mr. Guillot's introductions, and we esteem it highly from what we have seen of it. A beautiful, free flowering Rose, and very fragrant.

Madam Pernet-Ducher.

Hybrid Tea. A good, well formed bud, quite long and of a distinct shape and form. Color, a light canary yellow. The first distinct yellow Hybrid Tea. Good, strong, and of robust growth.

Madam Veuve Menier.

Hybrid Tea. A strong, vigorous grower, producing buds in great profusion. Pale, light rose; flower of good size, double, and very free flowering.

Madam A. Veysset.

"Striped La France." Hybrid Tea. A magnificent new Rose, identical in every respect with its parent, La France, except in two particulars—it is much stronger in growth, and the flowers are beautifully striped and shaded with a delicate white. The coloring is exquisite, and we think this Rose will please cut-flower buyers. It forces freely, and produces magnificent large buds and flowers.

Madam C. Testout.

Hybrid Tea. This is one of the introductions of '91. A really grand Rose, of the La France type, but larger and of better color. It is clear pink in color—there is nothing in the Rose line that can approach it in color, and the flower is as large as Baroness Rothschild and as free as La France, and if our judgment is not mistaken, it will make a sensation in the cut flower market when brought in in good shape.

Medea.

Tea. Flowers bright lemon yellow, with canary yellow centres. Very full buds, with high centres; foliage dark and thick, in color like Perle. This variety produces most of its buds of immense size, and very double. Texture firm and heavy, models of what a Tea Rose should be; on the other hand, quite a number of short, blunt buds are produced. It will make a good all round variety, we predict.

Edouard Littaye.

Tea. A fine, strong, vigorous Rose, after the general style of Madam Cusin. Flowering most abundantly; buds large, long, and of fine shape; full and double when open. Color, rosy carmine, tinted light pink, often shaded with violet pink. We are impressed with this Rose, and think, from its freedom of bloom, chaste color, and fine form, that it will be used for forcing.

Elise Heymann.

Tea. A vigorous, free flowering Rose, producing quantities of medium sized flowers. Color, light yellow, shaded with nankeen yellow, the centre rosy peach. A promising bedding Rose.

Bridesmaid.

A new and beautiful pink Rose of American origin. The buds are Mermet shaped and never come deformed. It is a vigorous grower and free bloomer, and said to excel the Mermet for its lasting qualities when cut.

Fifty Cent Marechal Neil Roses.

We have a large stock of nice plants, very strong, that we will keep dormant until late in Spring, so they can be shipped safely and begin to grow and flower as soon as planted out. These always give satisfaction. Price, 50 cents each.

Four Excellent Roses.

| PAPA GONTIER. | CATHERINE MERMET. |
| THE BRIDE. | PERLE DES JARDINS. |

In our estimation these four Roses have no superiors as ever bloomers. Perle des Jardins is the best yellow; Bride, the best white; Gontier, the best red; and Catherine Mermet, the best pink. Thus you have four distinct Roses, the best in their respective colors, that we can well recommend. Price for the four, one strong plant of each, $1.00.

Standard List of Roses.

These Roses are all grown in two and one-half inch pots, and are from four to eight inches in height; they are vigorous and thrifty, grown especially for our mailing trade. We would like to have our list of Roses carefully examined, as it is without doubt the finest in the country. State what varieties you have, if selection is left to us, and we will not send them. Many of the varieties enumerated in this list are new and of high merit, and sold last Spring for more than twice the price here named. There has been a number of new varieties introduced within the past few years of much value, but the price of them being high prevented their general distribution, and we now take pleasure in offering for the first time new and beautiful varieties at popular prices.

Price 10 cents each, 3 for 25 cents, 7 for 50 cents, 14 for $1.00, Purchaser's Selection of varieties. Our Selection from this list, all named, 16 for $1.00, by Mail or Express.

Aline Sisley. Violet rose; a fruity pleasant fragrance.

Arch Duke Charles. Brilliant crimson scarlet, shaded violet.

Antoine Verdier. Rich, dark, carmine pink, slightly shaded with white.

Adam. Bright flesh salmon rose, extra large size; double.

Adrianne Christopher. A lovely shade of apricot, citron and fawn.

Andre Schwartz. Beautiful crimson; a free flowering variety.

Aurora. Creamy white, shaded dark rose; very double.

Arch Duchess Isabella. White, shaded with rosy carmine.

Agrippina. Rich velvety crimson; few Roses are so rich.

Apolline. A clear pink, dashed with rosy crimson.

Alba Rosea. A beautiful creamy white with rose colored centre.

Anna Oliver. Very double; a creamy blush, shaded with deep carmine, and tinged silver rose.

Beauty of Stapleford. Deep rosy red, dark purplish rose centre.

Bourbon Queen. Clear rose color; large, double and sweet.

Bella. Pure snow white, splendid long, pointed buds; tea scent.

Bon Silene. A dark crimson rose, often changing to crimson.

Bougere. A bronzed pink, tinged with lilac; large and full.

Belle Fleur de Anjou. Beautiful silver rose with pointed buds.

Beau Carmine. Fine carmine red; very rich, good size, double.

Baron Alexandre de Virts. A delicate rose; highly perfumed.

Canary. Light canary yellow, beautiful buds and flowers.

Charles Rovolli. Lovely shade of brilliant carmine.

Clemont Nabonnand. A coppery rose, tinged purplish crimson.

Countesse Riza du Parc. Coppery rose, tinged with soft violet.

Catherine Mermet. Its buds are inimitable, faultless in form, and charming in their every shade of color, from the purest silvery rose to the exquisite combining of yellow and rose, which illumes the base of each petal.

Cels-Multiflora. Full and double, pale flesh, deepening to rose.

Cornelia Cook. The flowers are of the clearest, snowiest white, and arranged in the most faultless and symmetrical manner.

Coquette de Lyon. A fine yellow rose; large; not at all formal.

Clara Sylvain. Creamy white, good, full form, and fragrant

Comtesse de Barbantine. Flowers are large, beautifully cupped, full, sweet.

Crimson Bedder. Bright fiery velvety red; recommended.

Comte de Paris. Rich crimson, shaded bright purple.

Countess Frigneuse. Free bloomer; fine shape; a pure yellow; fine long buds.

Cheshunt Hybrid. Cherry carmine with a shade of violet; a good Spring bloomer.

Clothilde Soupert. Very double and beautifully imbricated, pearly white with rosy centre; free bloomer.

Countess Labarthe. One of the freest blooming Roses in existence.

Duchess of Albany. Rich deep pink; highly fragrant; also called the red La France.

Douglass. Dark cherry red, rich and velvety; large, full, double, fragrant.

Devoniensis. Magnolia Rose; beautiful creamy white.

Duchess of Edinburgh. Buds of the most intense deep crimson.

Dr. Pasteur. A handsome Rose, Hermosa pink, and very double.

Etoile de Lyon. Chrome yellow, deepening in the centre to pure golden yellow; flowers very large and double.

Esmeralde. A robust grower and free bloomer; beautiful silvery flesh with fawn shadings.

Ernst Metz. Rosy carmine, with salmon reflex; large pointed buds with long stems.

Gen. de Tartas. Rosy carmine, shaded purple; free bloomer; very sweet.

Glorie de Polyanthus. A bright rose with white centre; an abundant and fine bloomer.

Geo. Pernet. The prettiest of its class; silvery yellow, passing to peach rose.

Golden Gate. Buds long and pointed, having the beautiful form of Niphetos; color rich creamy-white, with centre and base of petals soft golden yellow.

Hermosa. Light pink; good bloomer.

Henry M. Stanley. A vigorous and healthy grower of neat, compact habit; quite full and very fragrant; color a clear pink, sometimes slightly tinged with salmon.

Isabella Sprunt. This Rose is a sport from Safrano, which it resembles in all respects save in color, which is a bright canary yellow.

Jean Pernet. A beautiful pale yellow, suffused with salmon; of medium size; beautiful buds.

Jean Ducher. Yellow shaded salmon; a strong and vigorous grower and profuse bloomer.

J. B. Varrone. A vigorous and healthy grower; bright pink, tinted salmon.

Joseph Metral. Cherry red with purplish tinge; a strong grower.

La Nankin. Apricot yellow; fragrant; good form; very distinct.

La Sylphide. Blush, with fawn centre; very large and double.

La Janquille. A saffron yellow; distinct and always in bloom.

Lauretta. A blush white, with peach centre, sometimes dotted with pink; very double and sweet.

La Princess Vera. A creamy white, bordered with coppery yellow; full and sweet; a good new Rose.

La Tulip. Creamy white, tinged with carmine; full and fragrant.

Lady Warrander. A pure white, sometimes shaded with rose.

L'Elegante. Vivid rose, yellow centre, shaded with white.

La Pactole. Elegant buds, color a pale sulphur yellow.

Louis Phillipe. A rich dark velvety crimson; free and beautiful.

Lucullus. A beautiful dark crimson maroon; full and fragrant.

Louis Richard. A coppery rose, beautifully tinted with lilac.

Louis de la Rive. A flesh white, inclining to a rose centre.

La Nuancee. Rose, tinged with fawn and coppery yellow.

La Chamoise. Nasturtium yellow; very beautiful buds.

Letty Coles. Rose color; large, full and globular.

Mignonette. A charming little Polyantha; a delicate rose, changing to blush.

Little Pet. Small white flowers; a free bloomer.

Luciole. Bright carmine rose, tinted and shaded saffron; the back of petals bronze.

Meteor. A clear bright red; beautiful foliage; of dwarf, vigorous habit.

Mad. de Watteville. White, slightly shaded salmon yellow, edged a bright pink.

Mad. F. Kruger. Orange yellow, shaded flesh; varies sometimes, according to soil; a good grower.

Mrs. Jos. Wilson. A beautiful Rose; deep cream color, edge of petals touched soft blush.

Mad. Philip Kuntz. Bright pink or china rose; fragrant, and a very free bloomer.

Mad. Scipion Cochet. A yellowish pink; very free bloomer.

M'me Augustine Guinnoisseau. The white La France. This is a very light variety of La France, and admired by many; we think it will make a good companion plant for La France and Albany.

Mad. Schwaller. Pink; large, fine; it flowers free and abundantly.

Mad. Berard. Apricot yellow; occasionally a golden yellow; large and double; of good substance and very sweet.

Mad. Celina Norcy. A delicate shade of rose; the backs of petals a purplish red; very large, full and of good habit.

Mad. la Conntess de Casnerta. Coppery red flowers; large; petals of good substance, but not full; splendid buds for bouquets.

Mad. Meline Vellermoz. A creamy white; thick petals, large and full, and slightly fragrant; an excellent variety for planting out.

Mad. H. Jamin. White centre, shaded yellow; large and full.

Marie Ducher. Vigorous and free growing; large, full flowers; color salmon; fawn centre; a splendid variety.

Mad. Joseph Schwartz. White, beautifully flushed with pink; of good size; cupped, and borne in clusters.

Mad. Louis Henry. Flowers medium to large, silvery white, shaded yellow; fragrant and of good form.

Moiret. Pink shaded salmon; very good Rose.

Mad. Bravy. Creamy white; large, full and very symmetrical.

Mad. Camille. A delicate rosy flesh, changing to salmon rose.

Mad. Caroline Kuster. A bright lemon yellow; very large.

Mad. Chedane Guinoiseau. A beautiful yellow Rose with fine long buds.

Mad. Margottin. A beautiful citron yellow, coppery centre.

Mad. Maurice Kuppenheim. Pale canary yellow, faintly tinged with pink.

Mad. Pauline Labonte. A salmon rose, large and full.

Mad. de Vatery. Red, shaded with salmon; of good form.

Marie Vau Houtte. A lovely pale yellow color, with outer petals most beautifully suffused with bright pink.

Mad. Hippolite Jamin. White, yellow centre, shaded pink.

Mad. Jure. Lilac rose, good size and substance; fragrant.

Mad'lle Rachel. A lovely Tea Rose; pure snow-white.

Marcelin Roda. Pale lemon yellow, with lovely buds and flowers.

Marie Sisley. A pale yellow, margined with bright rose.

Mad. Angele Jacquier. A light silvery rose, shaded yellow.

Mad. Falcot. Deep apricot yellow, with fine orange buds.

Mad. Dennis. Waxy white, centre fawn and flesh; large.

Mad. Dubroca. Delicate rose, shading to yellow.

Mad. Lambard. Rosy bronze, changing to crimson.

Mad. Welche. Pale yellow, sometimes cream, with short inner petals of glowing orange and copper.

Mad. Brest. Rosy red, shaded to crimson; large flowers.

Marie Guillet. Holds first place among white Tea Roses, in purity of color, in depth of petals, and in queenliness of shape.

Mad. Bosanquet. Flesh, shaded a deep rose; large size; sweet.

Mad. Damazine. Salmon rose, changing to amaranth.

Mon. Furtado. Yellow, well formed, full and fragrant.

Mad. Jean Sisley. Pure white; an elegant Rose.

Marshal Robert. White, centre shaded with flesh; large, full and globular.

Melville. Silvery pink; bright and elegant.

Niphetos. Large and very double white Rose of moderate growth; beautiful long pointed buds.

Princess Sagan. Rich velvety crimson; a free bloomer.

Princess Beatrice. Pale yellow; handsome glossy foliage; dwarf grower.

Princess Hohenzollern. Dazzling red; large, fine shape.

Perle des Jardins. Canary yellow; is large, full, well formed; very fine; this undoubtedly is the best dwarf yellow Rose in cultivation.

Perle de Lyon. Yellow, with salmon centre; large, full and very fragrant.

Purple China. Rich, purplish crimson; velvety.

Pink Daily. Light pink flowers, produced in clusters.

Premium de Charrisseans. A bright carmine rose, fawn centre.

Queen's Scarlet. A dazzling crimson scarlet, with beautiful buds.

Queen of Bourbons. A clear carmine, changing to clear rose.

Rainbow. A sport from Papa Gontier, originating in California; it has the shape of Papa Gontier, a lovely pink blotched and streaked with the darkest Papa Gontier color; to add to its great beauty the base of petals is of a rich amber.

Robusta. Clear carnation red, veined a rosy crimson.

Roi de Cramoise. A bright purplish crimson, full and very double.

Regulus. Brilliant carmine, with purple and rose shading.

Rosa Nabonnand. Imbricated, delicate rose, vivid in centre.

Rubens. A creamy white, with flesh centre; very large and full; superb.

Souv. Isabella Nabonnand. Large and globular buds of a charming light fawn and silvery pink

Souv. de la Malmaison. A noble Rose; the flower is extremely large, quartered and double to the centre; color a flesh white, clear and fresh.

Safrano. Bright apricot yellow, changing to orange and fawn.

Sombreuil. A beautiful white, tinged with delicate rose.

Souv. de Mad. Pernet. Beautiful, soft, silvery rose.

Sulphureaux. Sulphur yellow; fine in bud; fragrant.

Souv. d'Elise Vardon. Creamy white, delicately shaded with pink; fragrant.

Souv. d'un Amie. Rose, tinged with salmon; very large, full, highly perfumed; an old and reliable sort.

Souv. de Victor Hugo. Bright china rose, yellow centre, edge of petals carmine.

Snow Flake. Also called Marie Lambert; a free bloomer, in fact a white Agrippina.

Souv. Therese Level. Clear crimson, tinted purple; of distinct, fine odor.

The Queen. Pure white; good bloomer and of free growth; a white Souv. du Amie.

Triumph de Luxembourg. Rose carmine on a buff ground.

The Bride. The best pure ivory-white Tea Rose; the buds, which are of grand size, are carried high and erect on very bright, smooth stems.

Viscountess Folkstone. Exquisite in form of bud; color white with salmon shadings.

Waban. Resembles Catherine Mermet very much, except in color, which is a rich bright clear pink.

Valle de Chamounix. The back of the petals are a bright yellow, the centre highly colored with glowing copper and rose, every tint being clear and bright.

White Bon Silene. This is a sport from the old Bon Silene, differing only in color, being a pure pearly white.

White Daily. Pure white, beautiful long pointed buds.

Climbing Roses.

Noisette and Tea Scented.   15 Cents Each ; Strong Plants, 25 Cents.

America. Fawn yellow, changing to coppery yellow.

Anna Maria. Large; rosy pink.

Baltimore Belle. Pale blush; very double.

Bennett's Seedling. Pure white.

Beauty of Greenmount. Brilliant red; very hardy and vigorous.

Belle Lyonaise. Pale lemon yellow; extra fine.

Beauty de L'Europe. A deep yellow; reverse of petals coppery yellow; very large and full; free blooming and vigorous.

Celine Forestier. Deep sulphur yellow; flowers of good size.

Cloth of Gold, or Chromatella. Clear golden; large, full and double; highly fragrant; much prized for pillars and verandas.

Claire Carnot. Buff orange yellow, with peach blossom centre.

Caroline Goodrich, Or Running Gen. Jacqueminot. This is a hardy climbing Tea and should not be classed with the hardy climbers that bloom but once a year; the color is the same as that of Gen. Jacqueminot.

Climbing Devoniensis. The same as Devoniensis, except that it is a rampant climber.

Climbing Jules Margottin. Dark red.

Climbing Victor Verdier. Bright red' with purplish edged petals.

Dundee Rambler. White small flower; fine for rock work.

Estelle Pradell. Lovely pure white buds; flowers full and sweet.

Eva Corinne. A delicate flesh, changing to white.

Euphrouse. Lemon yellow, suffused with salmon.

Fortune's Double Yellow. Bronzed yellow or copper and fawn color.

Glorie de Dijon. A grand Rose; has many admirers; it is noted for the great size of its flowers, its delicate Tea scent and its exquisite shades of color, being a blending of amber, carmine and cream.

Golden Chain, Or Climbing Safrano. A beautiful pillar Rose, being a strong climber; color orange yellow or deep saffron; good size, full and sweet.

Gold of Ophir. Nasturtium yellow, suffused with coppery red; one of the most beautiful of climbing Roses.

Gem of the Prairies. Violet crimson.

James Sprunt. Deep velvety crimson; double.

Lamarque. Flowers of medium size, borne in large clusters; pure white and double; a most beautiful Rose.

Marechal Neil. A Rose so famous as to really need no description; its magnificent golden yellow buds are worn the world over, and floral work without Marechal Neil is usually regarded as lacking a proper finish.

Princess Stephanie. Salmon orange yellow, medium size, growth very vigorous.

M'lle Annette Murat. A seedling from Glorie de Dijon, with citron yellow flowers; of climbing habit, and very free flowering.

Marie Lavelley. Habit extra vigorous; flowers large and of fine form; color a vivid rose, shaded and lined with white.

Mad. Alfred Carriere. Extra large full flowers; very double and sweet; color rich creamy white, faintly tinged with pale yellow.

Musk Cluster. Creamy white, in large clusters.

Reine Marie Henrietta. A strong and vigorous grower; flowers large, full, of fine form, color a beautiful, pure cherry red; large, sweet scented.

Setina. Or Climbing Hermosa, Identical with Hermosa, except being of a vigorous climbing habit.

Solfaterre. Fine, clear sulphur yellow; good form, large, full and double; very sweet and good.

Waltham Queen. It is a strong grower and continuous bloomer; the flowers are large, full and sweet; the color is a rich scarlet crimson very beautiful; a profuse bloomer.

Washington. Medium size, pure white, very double; blooms profusely in large clusters; a strong grower and quite hardy.

Wm. Allen Richardson. Orange yellow; outer petals lighter; centre coppery yellow; very rich; rapidly becoming popular.

Hybrid Teas and Hybrid Perpetuals.

These Roses are all grown here in the open ground, and are large and vigorous. While they do not bloom with the freedom of the Tea Rose, there is nothing to compare with them for beauty and fragrance in May and June, and in the Fall.

Price 25 Cents Each, $2.50 Per Dozen.

Abel Carrier. A very dark crimson, with violet shade; centre a bright red; large, full and double.

Alfred de Rougemont. A pure white; double and lasting.

Annie de Diesbach. Bright rose; large and showy.

Alfred Colomb. Large, round flowers, very double and full; color clear cherry red, passing to bright, rich crimson.

Antoine Verdier. Flowers large and well formed, of a fine dark rose color, with a well defined line of silver on the edges of the petals.

Boule de Neige. A finely formed pure white Rose; occasionally shows light flesh when first opening.

Baron de Bonstetten. A clear velvety maroon, shaded with crimson.

Beauty of Stapleford. Dark purplish crimson; flowers well formed, large; a very beautiful and distinct variety.

Coquette des Alps. Large, white flower, occasionally shaded with pink; a constant bloomer, bearing its flowers in clusters.

Coquette des Blanches. One of the finest white; large, full and fragrant; fine for cemetery planting.

Chestnut Hybrid. Deep purplish crimson; fine flower, sweet scented, and inclined toward a climbing habit.

Caroline de Sansal. Pale flesh; large and full.

Duchess of Cannaught. Most distinct in foliage and blooms; a delicate silver rose with bright salmon centre; large and highly scented.

Earl of Pembroke. A bright velvety crimson.

Empress of India. A dark violet crimson; double and fragrant; a splendid Rose.

Francois Levet. Clear bright rose; a fine grower.

Gen. Washington. Brilliant rosy crimson; a good bloomer.

Gen. Jacqueminot. A velvety scarlet, changing to crimson; a free bloomer; good for Winter forcing.

Giant of Battles. A brilliant crimson; large, very double and sweet; esteemed one of the finest.

Hon. Geo. Bancroft. Flowers large, full and regular; bright rosy crimson, shaded with purple; very beautiful.

John Hopper. Bright carmine; good.

Jules Margotten. A bright cherry red; large, well formed, fragrant flowers; double and free; a splendid sort.

La Reine. A clear bright rose; of fine form.

Lord Bacon. Deep dark crimson, shaded with scarlet.

Louise Van Houtte. A rich crimson, heavily shaded with maroon; a beautifully formed and double flower.

Lady Emily Peel. A charming Rose medium size and full form; very sweet, color white, sometimes tinged and shaded with blush.

Marie Rudy. Vermilion red, shaded with crimson; large and full globular form; very fragrant.

Marie Baumann. An excellent Rose; very large, full and fragrant; color a rich, ruddy red, changing to a lovely scarlet maroon; very beautiful.

Mad. Alexandre Bernaise. Salmon rose petals sometimes edged with fine blush; full and fragrant; a very good variety.

Mad. Masson. Large and double; color reddish crimson; of fine form and substance; invaluable for bedding purposes.

Mad. Alfred de Rougement. A pure white, delicately shaded and tinged with rose; fully double, good size, very sweet and pretty.

Merveille de Lyon. Extra large, blush white.

La France. A peach shaded rose, that blooms through the Winter season; double and fragrant; we consider this the finest Rose grown.

Nancy Lee. A satiny rose; a delicate and lovely shade; of slender growth; flowers medium or small, and very fragrant.

Oxonion. A beautiful lilac rose, shaded with crimson; very large, double and sweet.

Prince Camile de Rohan. Very deep velvety crimson; large and full; a good Rose of splendid color.

Paul Neyron. Deep rose; very large and full; fragrant and free blooming; the largest variety known; very desirable.

Pride of Waltham. Delicate flesh color, richly shaded with bright rose; very distinct; flowers large and full.

Perfection des Blanches. A finely formed, pure white Rose; occasionally shows a light flesh when first opening.

Ulrich Bruner. Flowers large and full, with exceedingly large shell shaped petals; color a cherry red; a splendid variety.

Souv. de President Lincoln. Scarlet and crimson, shaded with a purplish vermilion; very full, fine form; beautiful and fragrant.

Fine Climbing Roses for the South.

The four following Roses are excellent for the climate of the South, where fine Climbing Roses are in demand. In order to place them within the reach of all, we will send the four by express for $1.00, one fine plant of each.

MARECHAL NIEL.   W. A. RICHARDSON.
CLIMBING PERLE.   CLIMBING NIPHETOS.

Climbing Niphetos.

There has never been but one point that held that glorious old Rose, Niphetos, in the background, and that was the serious defect of its being a weak grower, and this defect is now remedied. We have a vigorous climbing growth, coupled with all the charms of Niphetos. It has always held the lead as being the most elegant of white Roses. The buds on well-grown specimens measure four inches in length, and are produced in the greatest profusion.   Price 50 Cents Each.

CHRYSANTHEMUMS.

ROM almost pre-historic times the inhabitants of China and Japan have cultivated this famous flower. It is not many years since American florists have adopted it as a special object for their care, and during this brief time their progress has been so great that they have outstripped the older countries that had centuries of experience, and where Chrysanthemums were grown long before Columbus had thought of adding a new world to the old. There is not to be found in the annals of Floriculture to-day a parallel to the wonderful development of the Chrysanthemum under the influences of American environments. The climate of the South is in every way adapted to its proper development, as we have demonstrated here, and in all portions of the South they grow and bloom with wonderful satisfaction. We have seen blooms grown in the open air in Texas, Mississippi, North and South Carolina and Alabama that were as fine as can be produced under glass in the Northern States. For this reason the South should excel in the culture of this now all popular flower. Why is it necessary to go North to see fine Chrysanthemums when every facility for producing them can be found in the door yard of every cottager and artisan in the South? Let the past state of things be changed. Let every amateur and professional florist give a portion of their time to the grand development of the Chrysanthemum, and let the Northern growers henceforth turn their eyes Southward when fine Chrysanthemums are desired. It can be done, and it is not now too late to make a start. The future of the Chrysanthemum is yet long and brilliant; the meridian of its fame has not yet been reached. Its future may be predicted to this extent, that when distinctive forms no longer appear it will then have reached the meridian of its fame. From the new and striking varieties that each season brings into our plant commerce, it looks as if its resources had no ending, and consequently the great demand for them is likely to go on with renewed vigor for many years.

We have been eminently successful with the Chrysanthemum here, as have also the hundreds of growers throughout the South that have followed our instructions, and the collection that we offer has no equal in any catalogue of the present time. While our list may not be as lengthy as some, you may rest assured it contains all that is worthy of growing. If a customer should look through our list for a variety they want and cannot find it, they may feel assured that it was dropped from our list, and a variety of more recent introduction that has surpassed it is to be found in its place. The National Chrysanthemum Society of England catalogues nearly four thousand varieties at the present time in cultivation. What a bewildering list this would be for a person of only a limited knowledge of the varieties of Chrysanthemums to select a few dozen plants from. In order to simplify the task of selecting a few varieties, we have made our list of varieties as brief as it is possible to do, and retain all varieties of merit of recent introduction. Many old varieties are dropped from our lists forever, because they have been superseded by varieties of more recent introduction. This leaves a list to select from that must meet the approbation of all, as all are dropped that are not up to the standard of requirement of the present time and none are added but what has superior qualities to the older varieties in cultivation. The price is also within reach of all; we have only two lists of different prices, a ten and a twenty-five cent list, except a few novelties. We only offer this year what novelties we know to be good, as we visited the different shows of the North so we might see them and know what we were offering to our customers.

Cultural Notes.

WHEN TO PLANT.

The Chrysanthemum may be planted any time in the Spring as soon as frost is gone, and the ground can be worked, even up to the middle of June, and they will give good blooming plants in the Fall. The size of the plants is of little importance providing they are established, with good roots, and in a fresh and vigorous condition.

SOIL.

The Chrysanthemum thrives best in a compost composed of three parts of fibrous loam, one part well rotted cow manure, with the addition of about a six inch pot full of bone meal to a bushel of this compost. A handful or two of soot added to this will also keep it free from worms and add to the vigor of the plant. Dig the ground deep if cultivated in the open garden, and apply manure liberally. The Chrysanthemum is a gross feeder, and if well fed and properly thinned and pruned, the flowers will be large.

DISBUDDING AND STOPPING.

When your plant is six to eight inches high, cut back to a height of about four inches; allow four shoots to grow from this main trunk; when these four shoots are four inches high, pinch out the terminal bud, and save three or four of the branches that will spring from each of these limbs. Allow these last branches to attain a height of nine inches, then stop back for the last time. Allow one bud (flower) to a stem, and that a terminal bud.

WHEN TO LIFT.

In localities subject to heavy frosts in October and November, plants should be lifted into large pots or boxes by September 20th; after lifting them, drench thoroughly, and never allow them to suffer for water; give them manure water, if possible, once a week. If protected from frost by sheeting, they need not be brought into the house till well into October; and when brought in, place in a room without fire, and give plenty of air when not frosty. These instructions as to lifting apply chiefly to this State and further North. In all States South of this they do admirably in the open air. A covering of canvas is sometimes used to advantage when first Fall frosts appear.

MILDEW.

This is caused by cold nights succeeding sunny days, or two great extremes of temperature. Over crowding the plants and insufficient ventilation is another fertile cause of Mildew. Should Mildew actually appear, dusting the affected plants with powdered or flowers of sulphur is the best antidote, together with the maintenance of a dry atmosphere.

BLACK FLY.

The Black Fly or Aphis is the worst enemy the Chrysanthemum has to contend with; it infests the little plants in the early Spring, and will stick to them all the Summer long if not destroyed. A good decoction of tobacco water applied by means of a syringe or wisk broom, will make the plant so distasteful to them that they will soon forsake it. Soap suds from the laundry, applied in the same way as the tobacco water, will also help to drive them away. In sections where tobacco stems can be procured cheaply, if they are kept sprinkled among the plants the insects will never molest them.

SHADING.

If plants are grown in pots it is necessary to shade them through July and August. With a light canvas shading through the warmest part of the day they grow more vigorously and the labor of watering them is reduced considerably. It also keeps up a vigorous growth and the plants are consequently not apt to get dry and hardened as if fully exposed to the sun.

A FEW POINTS.

Don't expect large perfect flowers in poor soil from plants that have been dried up and neglected. Don't expect your blooms to be as fine as the catalogue cuts represent them unless you give them the same close attention as the blooms have had from which those cuts were made. Take any Chrysanthemum that you please, keep it growing and well fed, with plenty of drink; take the same variety and let it struggle along for existence, starved and thirsty. In the Fall cut a blossom from each plant and they will not resemble each other in any respect, not even in color. The same trouble will arise with a well grown plant if it is not carefully disbudded. Many a customer thinks his Chrysanthemum untrue to description, when the sole and entire cause of his disappointment is poor culture.

GENERAL COLLECTION OF CHRYSANTHEMUMS.

The varieties enumerated in this list are all good. All old varieties have been discarded to give place to newer sorts of higher merit. Only in cases of much merit have older varieties been retained. This list comprises all the novelties of two years ago, now offered at the following low prices.

Price 10 cents each, 3 for 25 cents, 7 for 50 cents, 15 for $1.00, Purchaser's Selection of varieties.   Our Selection from this list, all named, 18 for $1.00, by Mail or Express.

Advance. Incurved, of perfect shape, a deep pink, but quite distinct from every other kind.

Alice Bird. A large, compact flower of intensely bright buttercup yellow.

Alcyon. Very large carmine rose, striped white, reverse rose.

Ada Spaulding. A striking variety of globular form; petals very large, broad and solid; color shading from the purest pearl white to a deep pink on the lower petals.

Auriole. A lemon yellow, shading to white as it matures; flowers produced on stout stems; foliage fine; one of the best.

Aristine Anderson. Resembles Miss Mary Morgan; a decided pink, and larger blooms; double flowers with a high centre.

Alcazar. Incurved Japanese; of robust habit; flowers on strong stems; the upper surface of petals a bronze red, shading to yellow at tips, with reverse old gold.

Anna Dorner. Full, fine, bold flower, with outer petals striped and shaded a rich carmine, centre cream white.

Anna M. Weybrecht. Magnificent Chinese variety of purest snow white, the petals solid, broad and firm; of strong habit.

Anna J. Sprague. Flowers are very large, having widely spreading, reflexed petals; color white, lined and mottled with rosy pink.

Bolero. Rich chrome yellow; flowers large, reflexed; habit robust.

Beacon. Magnificent full and double creamy white flowers, borne on strong stems; the outer row of petals are tubular and reflexed, while those nearer the centre are incurved, with broad convex tips.

Ben d'Or. Handsome twisted yellow.

Christmas Eve. A magnificent white of great beauty.

Comte de Germiny. A bright yellow, with broad petals, shaded bronze.

Cullingfordii. A rich crimson shaded scarlet; very large reflexed flowers, beautiful and distinct.

Coronet. The richest golden orange, incurving to the centre; occasionally crimson stripes on the inner side of petals.

Cyrus H. McCormick. Very vigorous grower, producing stiff flower stems; resembles W. H. Lincoln in form; the flowers are large, a deep, dark yellow, shaded with bronze.

Clancey Lloyd. A delicate flesh pink, changing to pure white; petals medium in width, flat and cup shaped, incurving and covering centre.

Duchess of Manchester. Best white Chinese variety in cultivation.

Domination. A grand variety; flowers creamy white, large and fine.

Diana. One of the best whites.

Duchess. Deep red, tipped yellow.

Empress of India. A pure white sport from Queen of England; of same character.

E. G. Hill. Immense bloom of brightest golden yellow; full and very double; lower petals sometimes deeply shaded a bright carmine.

Evaleen Stein. In the way of Bride, but an improvement on that variety; of a delicate white.

Eda Prass. A fine, bold and recurving flower of great substance and depth; a lovely white, delicately shaded with blush.

Emma Dorner. Fine, deep violet pink, in the way of Violet Rose when finely done, but a deeper, purer color; large ball shaped flowers of splendid substance.

Etoile de Lyon. A grand variety; the color lilac, shaded silver, changing to white.

Ernst Asmus. Flowers of massive size and reflexed; the color is a rich chrome shaded amber, each petal having a line of puce down the centre.

Eldorado. Incurved deep yellow; of a dwarf, sturdy habit, with very strong flower stem; lasting a long time in flower.

Elmer D. Smith. The color is cardinal red, of a very rich and pleasing shade, faced upon the back of the petals with clear chamois; the flower attains great size.

Emily Dorner. The flower is nicely incurved, petals broad, and of richest shade of orange yellow, touched with crimson.

Esperanza. Flowers of good size; a blush, with creamy white centre; reflexed.

Earns. Rich red bronze; large, reflexed; fine habit; distinct.

Firenzi. A bright yellow, resembling Gloriosum; flowers large, on strong footstalks.

Flora McDonald. Most perfectly incurved, with very high centre; creamy white; good for all purposes.

Flora Hill. It is of splendid size and heavy texture; outer petals horizontal or slightly recurving; the creamy centre is perfectly full and incurved.

Frank Thompson. A splendid flower, very nearly spherical in form; petals very broad and heavy, finely incurving; it is very nearly white in color, only showing a touch of pearl pink at the base of the petals.

Ferd. W. Peck. Color a beautiful rosy pink; petals incurving, yet flower has an open appearance; perfectly double; very large.

G. F. Moseman. An improved bicolor, being a brighter color; bright crimson, tipped with golden yellow.

Gold. As name denotes, this is of the clearest golden yellow, and is perfectly double; one of the best new yellows.

Golden Plume. Very late; a clear and bright yellow; free bloomer.

Garnet. Showy Japanese variety; the inner side of petals a rich wine red, the reverse silvery pink.

Grandiflorum. A magnificent variety; petals very broad, incurving, so as to form a solid ball of the purest golden yellow.

Gloriosum. Very light lemon, with immense flowers, and having narrow petals most gracefully curved and twisted.

Harriet Beecher Stowe. Pure white petals, long, flat and pointed; very free flowering.

Harry E. Widener. Bright lemon yellow in color, without shadings; flowers large, on stiff, stout stems; flowers are erect, without support; incurving, and forming a large rounded surface; the petals crisp and stiff.

H. Waterer. Late yellow, with reflexed petals.

Innocence. Seedling from Mrs. Hardy and as fine in form, texture and finish as the parent, but without the velvety pile.

Ithaca. Outside petals tubular, incurving, so as to form a perfect ball; color white, shaded pink.

Ivory. One of the finest whites now in cultivation; very large, globular and early.

J. C. Vaughan. Richest plum crimson without any shade of purple; flowers reflexed; very large, strong and stiff stems.

John Firth. Petals cup shaped, and arranged in compact rows one above another, completely covering the centre; deep Mermet pink, shading to silvery rose.

John Goode. The outer petals are of a delicate lavender, forming a decided band of color, the inner petals are clear lemon; a plant in bloom has a most beautiful, airy appearance.

Jessica. Snow white, with yellow centre; very large flowers.

John Thorpe. Color the richest deep lake, a new shade; very vigorous and early.

Kioto. A beautiful incurved yellow of fine form and habit.

Kate Rambo. Pure white; a very large, broad, double flower; florets curl at the tips and slightly incurve; fine, robust, yet compact grower, with strong flower stems.

Lily Bates. Very large and perfectly double, clear, bright, rich pink; the petals broad and flat; a new and distinct form.

Lizzie Cartledge. Very large, full and double flower, incurved, except the under row of florets, which reflex; the color a bright dark rose, reverse silvery white.

La Tonkin. Centre white, shaded rose on the outside; flowers large.

L. Canning. A most exquisite white, absolutely pure; the flower regular in form; very large and flat petals.

Lillian B. Bird. Of the very largest size, with full, high centre; petals tubular, of varying lengths, the flower when fully open being an immense half globe; the color an exquisite shade of "shrimp pink."

Lucrece. A pure white, resembling Christmas Eve, but surpassing that in size, form and lateness; largely used for cutting and late decorations.

Louis Boehmer. This is also known as the pink ostrich plume; it has the wonderful hair like growth of Mrs. A. Hardy; the blooms are beautifully incurved, inside petals of a deep rose, shading to a lavender pink on outer edge.

Mrs. Mary Morgan. Rich, deep pink; perfect shape; incurved.

Mountain of Snow. Good white; very large.

Mad. Thibaut. Fine dark red; valuable as a late variety.

Mrs. Geo. Bullock. A pearly white; flowers very large and flat; very fine for exhibition purposes.

Mrs. Irving Clark. Pearl white in the margin, shading to deep rose in centre, which is beautifully whorled.

Mad. Domage. Early white.

Mrs. Richard Elliott. Another grand yellow; very large, slightly recurved; petals long, of medium width.

Mrs. A. Blanc. Centre of floret erect; outer petals horizontal or drooping, of rosy lavender, centre soft clear rose.

Mrs. Vanaman. Cherry red; very large and perfectly distinct.

Mr. H. Cannell. Very rich and broad petals; color of brightest possible yellow.

Mrs. Fottler. Large, full, double flowers of clear, soft rose, the shade of the La France Rose.

Mrs. E. W. Clarke. Deep purple amaranth, silvery rose reverse; very large and highly scented.

Mrs. Jessie Barr. A fine incurved pure white of large size, with flat, ribbonlike petals.

Mrs. Langtry. An enormous incurved Japanese; the outer petals long and quilled, the inside ones flat and most beautifully incurved; the color a pure white.

Molly Bawn. A most valuable white variety, for its size, shape and purity.

Mrs. C. H. Wheeler. Flowers are of the largest size, and of heavy substance; color a bright crimson on the upper side of petals, while the under side is clear old gold.

Mrs. Frank Thompson. Very large incurved, with broad petals; mottled deep pink, with silvery black; very distinct.

Mad. Louise Leroy. White, tinted a blush; very free blooming, and rather late.

Mrs. Winthrop Sargeant. A brilliant straw color, incurved; flowers borne on long, stiff stems; very large, if not the largest in this line of color.

M. P. Mills. In shape it somewhat resembles a mushroom, and has very thick flowers of great substance; color an orange yellow, sometimes faintly streaked with red.

Mrs. L. P. Morton. Bright pink, base of each petal pure white.

Miss Mary Wheeler. Pearly white; of immense size and very double; the petals delicately tinted on edges a pale pink.

Mad. Baeo. An extra large, deep rose, tipped golden; very double; a magnificent show bloom.

Miss Minnie Wanamaker. A creamy white, incurving from first opening to finish, when it appears as a snow white ball; is rather dwarf in habit.

Mrs. J. N. Gerard. Grand cup shaped variety, closely incurving with age; of large size, and of the brightest peach pink.

Mrs. President Harrison. The largest of all the Mrs. Wheeler type, on which it is an improvement, both in constitution, size, color and habit.

Mrs. Andrew Carnegie. Bright deep crimson, reverse of the petals a shade lighter, broad, long and flat; of feathery texture, with strong, erect, heavy footstalks.

Mrs. Alpheus Hardy. Flowers pure white; on the upper surface of the floret petals is what at first sight appears hoar frost or snow, which gives it a chaste appearance.

Mrs. Libbie Allen. Of splendid form; the color is a yellow of a very pleasing shade; a good variety for exhibition blooms.

Mrs. John Westcott. Color cream pink to cream white; reflexed, with very stout petals of the most symetrical form.

Mary Waterer. Flowers of a delicate rose shade, of immense size; very attractive, durable and double; of short but very healthy growth, and has good stiff stems.

Mrs. R. D'Oyly Carte. Flowers of a medium size, incurving so as not to show the centre when fully expanded; this is one of the purest pink shades yet seen.

Mrs. Oliver Laughton. A very large, finely incurved flower; the inner part of the petals are rosy purple, with rose shadings.

Mrs. A. Rogers. A rich golden yellow; incurved; form of flower perfect; each bloom a bouquet; has produced flowers over nine inches in diameter.

Mrs. Herbert A. Pennock. Flowers large and of a beautiful orange yellow color; it has very strong stems, bearing its enormous flowers perfectly erect.

Mrs. D. D. Farson. Size immense, solid and compact; color a bright Mernet pink.

Mrs. Lay. A chaste and very beautiful large incurved flower; petals are cup shaped, white with faintest blush lines on edges.

Mrs. Kendal. Color rich Jacqueminot with reverse of petals copper bronze, shading to gold from base to tips; free bloomer of good habit.

Mrs. I. D. Saller. A flower of largest size, finely incurving, with broad and sharply pointed petals; strong grower, producing heavy flower stems; color a soft shell-pink.

Mattie C. Stewart. Clear bright golden yellow; extra large and double; the petals broad and flat, reflexing with age.

Mistletoe. Of the Comte de Germiny type, with the outside of the petals silvery white, lined within with crimson; has wide concave petals.

Mermaid. A very delicate yet bright pink, perfectly incurving and broad petals; extremely delicate in color and finish.

Mattie Bruce. Silvery pink in color; of medium size.

Mrs. R. J. Bayles. Immense incurving Japanese bloom; color a clear yellow, striped and highly marked red, bronze and old gold.

Nymphæa. A new sweet scented variety, the flowers resembling a Water Lily, hence its name; has a most pleasing fragrance.

Potomac. Bronze, red and gold flowers, reflexed, of medium size; very bushy habits.

President Harrison. Immense cupped flower, the outside petals salmon red, centre deep Indian red.

Potter Palmer. Color pure white; flat on opening, but gradually assuming a half globular form of great size.

Progression. Extra large late flowering variety, and remaining in flower up to Christmas; color purest white; very double.

Philip Breitmeyer. A most distinct variety, having heavy stems and foliage of a beautiful yellowish green; the flower is of the brightest golden yellow.

Penelope. Flowers large and globular; the petals are broad and nearly erect, the color being white, clouded with a beautiful deep rose, and tipped with buff.

Pandanus. Strong stems, free grower; pure white petals of good substance; a perfectly double flower of very large size.

Puritan. White, tinted with rose, large, good habit, and one of the finest for bush plants.

President Arthur. Immense rose colored flowers, opening in whorls.

Prince Kamoutski. Large, incurved, of the Comte de Germiny type; inside of petals bright crimson, outer ones a deep coppery bronze.

Pelican. The finest of recent introduction; pure white shaded cream; petals irregular, flat, half tubular.

President Spaulding. Very vigorous; of a rich reddish purple color.

Robert Bottomly. Pure white; flowers large; very fine.

Robert Flowerday. The outer petals flat with centre incurving, upper surface a bright crimson luke; reverse silvery pink.

Rohallion. Petals long, reflexed and twisted; color a dark yellow; a strong grower.

Russell. Flowers very full, reflexed; snowy white; of good size; habit of the plant free and robust.

Robt. S. Brown. Will make a magnificent exhibition variety, either as a cut flower or grown in pots; color richest crimson.

Rose Queen. A self color of bright rose amaranth; very distinct shade; flowers broad cup shaped; five inches across; one row of petals quilled half way and then flattened.

Superbe Flore. Globular flowers; a carmine rose; lighter centre; white reverse.

Sallie McClelland. Beautiful flowers, faintly shaded; a blush white; of fine habit.

Salvator. Flowers large, of a rich crimson red shade; reflexed; very robust and quite distinct.

Shasta. Long pure white tubular or needle-like petals; dwarf habit; a decided improvement on Mrs. Cleveland.

Sunnyside. A delicate flesh tint while opening, becoming white when fully expanded.

Sugar Loaf. The color is varying shades of yellow, often shaded bronze, and sometimes perfectly clear; a giant grower.

The Bride. This has been placed at the head of all white varieties.

T. C. Price. A perfectly double flower, much twisted and incurved in form of a corkscrew.

Triumph. A splendid and well formed flower; full centre, with broad, heavy, finely reflexed petals; color a beautiful deep rose-shaded amaranth, the under side of petals light pink; plant a strong grower.

Thos. H. Spaulding. A very brilliant and effectively colored variety; flowers very large, of a rich red crimson, with golden yellow tips, partially incurved; fine healthy grower.

Uji. The most novel colored flower yet seen; the ground color is a deep drab brown, overlaid with crimson lines; unique in effect.

Vivian Morrell. Large pink and of fine form; was among all the prize sets at the various shows last Fall.

V. H. Hallock. Rosy pearl, of a waxy texture; the flowers are six inches in diameter; the petals are convex, and rounded downwards half their length from centre, changing to a beautiful curved form.

Violet Rose. A grand double variety; exceedingly free; of perfect form; beautiful combination of violet and rose in color.

Wm. H. Lincoln. A magnificent golden yellow variety, with straight and flat spreading petals; an extra large flower, completely double.

Walter W. Coles. Very bright reddish terra cotta, reverse pale yellow; outer petals long, broad pointed and horizontal.

Yonitza. Chinese incurved of the most perfect form; flowers stand erect on stout stems; the outer petals droop as they mature, forming a perfect ball; color white, with a delicate shade of green.

Waban. Pink; very large and full flower; long and broad petals, the outer of which reflex; those of the centre incurve, making a superb and beautiful show bloom.

Wootten. This is a good variety, either for pot specimens or for cut blooms; the flowers are pure white, large and globular.

New Chrysanthemums of 1892.

The following list comprises all the novelties of last year, and contains many of the finest Chrysanthemums in cultivation at the present time. We bloomed them all here last Fall, and can highly recommend them. These plants sold last Spring at $1.00 and $1.50 each, and we offer them here for the first time at a price that is within the reach of all.

Price 25 cents each, 3 for 50 cents, 7 for $1.00, 16 for $2.00. Our Selection from this list $1.50 per dozen.

Ada H. LeRoy. A fine symetrically formed variety; petals broad and cupped full to the centre; color deep rose pink; extra large flower; one of the best for exhibition purposes.

A. Ladenberg. A full and extremely double Japanese variety of immense size, having been grown ten inches across; style and shape of Mrs. Irving Clark, except in color, which is a clear delicate rose pink; magnificent for exhibition specimens.

Col. Wm. B. Smith. Immense double high built flower; petals very broad and large, forming a solid mass of the richest bright golden bronze.

Clara Bertemann. Old gold; sport of John Collins, with flowers of the same shape; opens at first a bronze color, with purplish stripes underneath, and then changes to a beautiful golden yellow.

C. B. Whitnall. Large and very double Chinese incurved variety; of a velvety maroon, lined in a shade lighter; first approach to this color in the globular Chinese.

Col. H. M. Boles. Vigorous in habit; very large, rose-pink flowers with twisted petals, veined with a lighter shade; very full centre; a grand exhibition variety.

David Rose. Rosy claret, edged with silvery white; large and handsome.

Dr. Covert. Large golden yellow, deep and bright; heavy in texture, very double and incurving, stems strong; an improved Mrs. W. K. Harris in size and color; follows Widener in season of blooming.

Dr. Callandrean. A creamy yellow, shading to white; petals very large, broad, and inclined to incurve; a very beautiful blossom.

Dr. H. A. Mandeville. Large, full and double, bright chrome yellow; petals very long and twisted, with a swirled habit; the upper petals incurve, while the lower ones reflex; one of the grandest varieties in cultivation.

Evening Glow. Deep yellow in centre, with rich bronze red shadings toward the ends of each petal; very much like the coloring of a brilliant sunset.

E. A. Wood. Incurving and very fine shaped flowers; velvety crimson, reflex of petals shaded gold; lower petals slightly reflexed, which gives it a beautiful appearance; the foliage is very heavy, dark green; it has strong, stiff stems, and is a good grower.

Eva Hoyte. An immense double Japanese bloom of clearest and brightest yellow, a solid ball with full and high built centre; grand for any purpose, and superior to Widener or other existing varieties for exhibition. Two hundred and fifty dollars was paid for the control of this magnificent yellow variety.

E. Hitzeroth. Magnificent extra large flower; the petals broad and peculiarly arranged, completely filling the centre; a bright lemon yellow, exquisite and novel form; superior for commercial or exhibition purposes.

Ella May. Primrose yellow, apricot centre; extra large and fine.

Edward Hatch. An immense variety, incurving grandly, in globe form with reflexed outer petals; color soft lemon and bright pink; one of the grandest of the year.

Exquisite. Delicate pink; a very fine full headed flower; petals rather broad and standing pretty well out all over.

Faultless. Deep golden yellow; extra large and full; in color and form unequalled; measured over eleven inches in diameter.

Fred. Dorner. Opens with an incurving centre, and reflexed outer petals, but gradually assumes a pyramidal form of pure white, faintly penciled in light lavender; flowers large, on stiff stems, and a good grower.

Goldfinch. Deep, intense crimson on the upper surface, the reverse being a light bronze; a pleasing contrast.

Golden Gate. Imported Japanese, 1890. Tawney yellow, quite distinct, full centre; large spreading flowers, slightly whorled, regular form; stout stems; long white petals of good substance; a good cut flower variety.

George Savage. Flowers very large; a pure white, with broad, strongly incurved petals, making the flower almost hemispherical and very solid; a grand variety of vigorous, free flowering habit, and useful for cut blooms for exhibition.

Goguac. This variety needs but little description, as we are all familiar with Mrs. Irvin Clark, from which it is a sport, and identical in every way except in color, which is white; one of the best sports of late years for florists' use.

Harry Balsley. A fine cut flower variety; we predict this novelty will be held in high esteem for the beautiful shade of pink it possesses, which is a pearl pink shading to a Mermet pink; flowers are double, semi-globular, with erect petals.

H. F. Spaulding. A grand Japanese variety of most novel shape and effect; color rich apricot yellow shading to rose, centre petals clear yellow, bloom solid and double, high built, and of the largest size, is similar in shape to a pineapple; habit very strong and robust.

Harvest Moon. Clear, bright yellow; petals flowing and somewhat twisted; foliage clean and plentiful; a beautiful variety.

Hazel Gallagher. Large and incurving flowers; bright rose-pink, the reflex of petals silvery pink; the flowers form a complete ball; very distinct; stiff stems and a good grower.

Harry May. Flower very large, double, and forming a massive sphere; the color deep old gold, occasionally veined red; petals broad and thick; a magnificent prize winner.

Indian Chief. The color is a vivid and strong crimson; the flower very large, loosely incurved, and decidedly Japanese in form.

Joseph H. White. A large white variety with wide open petals of Dahlia-like form; a nicely rounded flower of fine size and substance, and perfect purity of color; a fine grower; good stiff stem.

John Bertermann. A strong and clean growing variety, with immense white flowers of a rather flat form, and very double; the centre a creamy white; the petals quite thick and lightly corrugated; a grand variety; will require a very little heat to bring it in for the shows.

J. N. May. Extra large deep ox-blood red, the color of Mrs. J. T. Emlin, but much larger and more double; reverse of petals shaded a copper bronze; full and solid bloom; a magnificent acquisition.

Julius Roehrs. Rich violet rose, the reverse silvery pink; a unique and charming contrast; flowers incurved, of largest size.

John H. Taylor. Large and reflexed plumed flowers; ground of the flower white, flaked and mottled a lovely shade of pink; strong stems; growth of the best.

J. Schuyler Mathews. Style of Orchard; finely incurved, showing no centre; red bronze, reflex of petals gold; one of the finest of its class; strong stems; good grower.

King's Daughter. Imported Japanese, 1891. Pure snow white, very long and drooping outer petals; centre petals irregularly incurved; stout stems, very showy, loose, pure Japanese style.

Ki Ku. One of the most peculiar Chrysanthemums in cultivation; the lower petals are broad and flat; color a deep pink, centre petals silvery pink; long and tubular, which gives the flower a peculiar shape; large flowers and very strong stem; very distinct and a new form.

Lillian Russell. Beautiful and broad petals of a clear silvery pink, incurving and forming an immense round ball of the very largest size; an early and fine flowering variety, suitable for all purposes.

Mrs. Maria Simpson. A most perfect incurving Japanese; the petals very broad and channeled, and closely incurving so as to completely cover the centre; a fine chrome yellow of largest size.

Mrs. A. J. Drexel. A large flowering early variety of fine crimson lake color; form rounded and beautiful and half globular; too early to be seen at its best at the shows.

Mrs. J. W. Morrissey. A mammoth flower with full and double centre; it was exhibited twelve inches across; color a silvery pink, the inner surface of petals bright rose; grand exhibition bloom.

Mrs. F. Schuchardt. A charming variety, which, like all in this collection, is of the largest size; coloring exquisite, the centre being creamy white, with the ends of the petals tinted a delicate rose.

Mrs. I. Forsterman. A magnificent extra large snowy white variety of the Japanese incurved type; superb grower and free bloomer.

Mr. D. S. Brown. Flower of medium size, semi-double, of a clear canary yellow when first opening, but changing to cream as the petals expand; distinct color.

Mrs. W. H. Phipps. A grand variety, after style of Domination; pure white, large massive flowers; strong and stiff stems; fine grower; it is later than Domination, which will make it very valuable.

Miss Heylett. Flowers a very pleasing shade of violet-amaranth, slightly incurving; reverse of petals shaded much lighter, after the style of Mrs. E. W. Clarke.

Mrs. Georgie Cole. Imported Japanese, 1891. Garnet purple, reverse of the petals lighter; large glittering flowers, very double and reflexed; inner petals slightly whorled; distinct in color and shape.

Mount Whitney. An exceedingly large, pure white flower, quilled, and of a somewhat spreading, flat disposition; splendid.

Mrs. John Eyerman. Flower is semi-globular, full double, petals decidedly spoon shaped, the lower half being tubular, while the limb is broadly expanded; upper surface rose pink, lower pale lilac; a grand variety for exhibition purposes.

Mrs. Senator Hearst. Pure white; a full, flowing flower; petals spreading, a little twisted and irregular, and of different shapes and sizes, many thread like ones being intermixed among the larger ribbons.

Mrs. Gallagher. Large, semi-globular bloom, with full centre; deep maroon crimson above, paler below; one of the very best dark varieties.

Mrs. J. Hood Wright. Flowers large, full double, of the purest white, with reflexed, twisted petals; strong grower, and one of the finest early varieties in cultivation.

Mrs. Robert Craig. One of the finest white varieties in existence; in form the purest Chinese incurved, while in size and texture and general build it has the grandeur of the Japanese; color snow white.

Mrs. L. C. Maderia. A perfect, compact globe of bright orange yellow; the petals upright, of heavy texture, and like unopened quills; of strong constitution; stems stiff, and flowers of large size; color, form and texture absolute perfection.

Marguerite Jeffords. A large ball of fine amber color; a splendid grower and one of the notabilities of the past year.

Majesty. Deep, glowing red; similar to Cullingfordii, but larger and finer; the color being more intense than that fine variety.

Miss M. Colgate. Flowers perfectly hemispherical, compact, full centre, pure white; petals broad and incurved; a grand variety for exhibition purposes.

Mr. Hicks Arnold. A strong growing and floriferous variety, bearing large, full double flowers of an old gold color, spherical in shape.

Miss Annie Manda. Flower high and compact, perfectly double, incurved, of the purest white; the petals are covered with long hair-like growths, giving it a unique appearance, surpassing Mrs. A. Hardy in form and habit, being sweetly scented.

Mrs. E. D. Adams. Very large; petals of medium width, outer ones swirled, as if the flower had been turned swiftly on its stem; color pure white; specimens measured eighteen inches from tip to tip.

Miss Ada McVicker. A plant of strong habit, producing immense creamy-white flowers, with broad, thick, reflexed petals; a grand variety, and one of the best for specimen blooms or other purposes.

Mrs. W. S. Kimball. Flower very large, full double, pale blush or creamy white with a yellowish centre; the petals are very broad and reflexed; one of the very finest varieties for exhibition purposes.

Mrs. Jerome Jones. Pure white, incurved, high round flower, the true highest type of best Chrysanthemum flower to date. A close observer at the New York, Philadelphia and Boston shows says: "I consider it the best white shown in 1891."

Mrs. Governor Fifer. Japanese. The flowers flat, with pure white, broad petals; centre incurving when fully matured; double at all stages of development; stout stem, with fine foliage; was shown at Indianapolis eight inches in diameter.

Mrs. C. D. Avery. Japanese. Petals long, convexed and twisted at maturity; color a new shade, pure dandolin yellow, darker than Lincoln or Widener, without any bronze or red shadings; perfectly double; strong stems; clothed with very glossy dark green foliage; we can recommend it as first-class.

Olga. Pink; it is a fine and well-built flower on stiff stem, full and double to the centre; color soft rose, with magenta shadings; it received first prize in competition with all the crack seedlings of the year grown by experts.

Oliver Wendell Holmes. Slightly incurved; light mahogany-red, reverse of petals light yellow; flowers rather flat; long, stiff stems; fine grower; one of the best of its class.

O. P. Bassett. The finest red Chrysanthemum ever offered; the color is identical with that of Cullingfordii, while the size, breadth of petals and general build are grand; is a fine grower, with large, leathery foliage.

Patrick Barry. The new golden yellow hairy Chrysanthemum; flowers very large, of the very brightest golden shade; habit vigorous and tall; stems stiff and erect; this is the grandest introduction from Japan during the past year.

Popularity. Delicate flesh pink, finely incurved, large and handsome; the ends of the petals are cut so as to resemble a stag's antlers.

Ruth Cleveland. A chaste and beautiful acquisition of largest size; broad and cup shaped, the outer rows reflex, inner ones incurved, forming a high built centre of most delicate silvery pink. This variety being registered, is the only one recognized by the American Chrysanthemum Society as being entitled to this name.

Sweet Lavender. White, shaded with blush, lower petals tinged with lavender; flowers of large size and form, slightly perfumed, from which fact it derives its name.

Secretary Furson. This is a large yellow ball with tubular petals, and in general appearance like Louis Childs Maderia, but larger in size and clearer yellow.

Spartel. This is another addition to the Hardy type; color delicate rose-pink without any trace of violet or magenta; the hairy-like filiments are very prominent; constitution, foliage and general growth resembles Fabre, which is perfect.

Surprise. A unique shade of bronzy red, bold, handsome flowers of the very largest size, and of fine globular form.

Volcanic. Flowers are reflexed and plumed, of a lovely canary-yellow, the reverse of petals lighter; as the flowers open they reflex to the stem, with the centre lighter than the body; this gives it a beautiful appearance; no other yellow Chrysanthemum is of this peculiar shade; strong stems and a good grower

Chrysanthemum Novelties

You Cannot Afford to Pass.

The following are of perfect habit, perfect flower, perfect color, and perfection in every respect.

Roslyn.

A superb clear Mermet rose pink, petals thick and heavy, cup shaped, solid to the centre; immense in size, having been exhibited eleven inches across; habit the best; stems stiff and erect, covered with the most luxuriant foliage; the best pink in commerce. Awarded silver medal by the Pennsylvania Horticultural Society, Certificate of Merit at the Madison Square Garden Exhibition, and was one of the varieties winning the Spaulding Prize at Philadelphia, Pa., for the best six new seedlings. Price 35 cents each.

Golden Wedding.

Richest golden yellow, intense and dazzling in color; flowers ten to twelve inches in diameter, four to six in depth; petals broad and long, double to the centre without an eye; the grandest of yellows. Awarded silver medal at Philadelphia, Pa., 1892; the Cutting Cup, Garden and Forest Cup, and the first premium at the Madison Square Garden Exhibition, 1892. A prize winner wherever it has been exhibited. Price 50 cents each.

Tuxedo.

A beautiful reflexed variety; dwarf grower, with handsome foliage; the color is a clear amber, and without doubt the best in this beautiful color; it is good either for specimen pot plants or exhibition blooms. Price 25 cents each.

Zambesi.

"The new Yellow Hairy Chrysanthemum." This variety resembles W. A. Manda in its hairy petals, yet it is not synonymous. It partakes of Lincoln in color (bright yellow) and dwarf short-jointed growth. The foliage is similar to that of Hardy, only much larger. It is of dwarf growth, for which everyone admires it very much, and makes a pretty pot plant. Price 25 cents each.

Shenandoah.

Magnificent broad flower of immense size, full and double to the centre; petals over an inch in width; color a new shade of deep chestnut brown on both upper and lower surfaces, the entire bloom being a solid color without shadings; novel and distinct. Philadelphia prize winner. Price 25 cents each.

George W. Childs.

"The Jacqueminot Crimson Chrysanthemum." No Chrysanthemum created such a sensation as did this, when exhibited at Orange, N. J., and Philadelphia.

The flowers are of massive size, reflexed, with broad, stiff petals; color deep, rich, velvety crimson, with no shade of brown or chestnut. Flowers borne on strong, stiff stems. The foliage resembles Cullingfordii, but is much heavier and darker. They grow close up to the flowers, which gives the plant an elegant appearance. It is a fine, strong, clear grower. It was exhited at Indianapolis this year over eight inches in diameter. No Chrysanthemum can compare with it for color, size of flower and growth. Price 35 cents each.

Winnie Davis.

This new seedling was raised by us here and was exhibited by us during our display in Nashville, where it was named by the ladies in charge in honor of Miss Winnie Davis, the daughter of the Confederacy. It is a vigorous grower and free bloomer; has large spreading blossoms when disbudded. The petals are tubular in form, pure white at the base, and deepening to a rich lavender towards the ends. When fully matured the color of the bloom very nearly resembles the true Confederate grey. Nice strong plants, 25 cents each.

W. A. Manda.

"The new Golden Yellow Hairy Chrysanthemum." Flower very large, of a clear golden yellow color; the plant is vigorous in growth and the flower is borne upright on a stout stem; this is the grandest introduction from Japan made during the past year, and no collection can afford to be without this variety. Price 25 cents each.

Niveus.

A grand snow white variety; centre irregularly incurving, with outer petals reflexing nearly to the stem. Constitution robust; foliage large and abundant, while the keeping qualities of the flowers are unsurpassed. Without doubt the best seedling of the year. Winner of the gold medal and $100 at Cincinnati; the Periam & Thorpe Special Prizes at Chicago. Also $25 at Indianapolis, and the same at Bay City, as the best seedling, any color, for 1892. Also Certificate at Springfield. John Thorpe says: "Niveus, I predict, will not be deposed for eight years, as it has in its composition nearly everything that at this day leads me to believe is nearly perfection; its snowy whiteness is beyond cavil, its elegant shape is such as to please the most fastidious, its size is far above the average without the slightest approach to coarseness, its footstalks are strong and graceful, its foliage is perfect, in other words it is a gem." Price $1.00 each.

Selections for Large Cut Blooms.

The set of twelve for $3.00.

C. H. McCormick.	Marguerite Jeffords.	Golden Wedding.
Roslyn.	Lizzie Cartledge.	Mrs. Gov. Fifer.
G. W. Childs.	Eda Prass.	Tuxedo.
L. B. Bird.	Golden Gate.	A. Ladenburg.

The following set of twelve with the above would make an excellent list of twenty-four varieties for specimen blooms.

This set of twelve for $1.00.

Exquisite.	Miss Minnie Wanamaker.	Potter Palmer.
Frank Thompson.	Louis Bochmer.	Jno. Berterman.
V. H. Halock.	Lucrece.	Mrs. Libbie Allen.
Vivian Morell.	Mermaid.	Mrs. W. Bowen.

Here are Thirty Fine Varieties to Grow in Pots for Specimen Plants.

One each of the thirty varieties for $3.00.

Whites.

Pelican.	Jessica.	L'Canning.
Sunnyside.	Ivory.	Minnie Wanamaker.

Yellows.

Alice Bird.	Dr. Covert.	Sugar Loaf.
Wm. H. Lincoln.	Kioto.	Zambesi.

Shades of Bronze.

G. F. Moseman.	W. W. Coles.	Edwin Molyneux.
Alcazar.	Prince Kamontski.	C. H. McCormick.

Shades of Pink.

Alcyon.	Lillian Russell.	Mattie Bruce.
L. B. Bird.	V. H. Hallock.	Frank Thompson.

Shades of Red.

G. W. Childs.	C. B. Whitnall.	President Harrison.
Robt. Flowerday.	Cullingfordii.	Marguerite Jeffords.

A Special Offer to Those About to Form Collections.

One hundred varieties, our selection, containing a fine assortment from both lists, including Novelties, for $10.00. One-half of each for $5.00.

Other Collections.

One hundred distinct, choice and rare kinds, our selection, $5.00.
Fifty distinct, choice and rare kinds, our selection, $3.00.
Fifty distinct kinds, two of each, one hundred plants in all, our selection, $5.00.
One dozen extra choice varieties $1.00, free by mail.

Prices on Application.

We have in stock many varieties not quoted in this Catalogue, and request those who desire other kinds to mail us a list of their wants. Prices will be quoted on application on all Novelties of 1893 not herein enumerated. We are the largest growers of Chrysanthemums in this country, and are thoroughly acquainted with the habits and qualities of every variety we grow, so that parties not familiar with the different varieties would do well to leave the selection of varieties to us, only stating for what purpose they are required, when we can make a better selection for them than probably they could themselves. Our Chrysanthemums have been prize winners everywhere we exhibited them, in the North as well as the South. We would be glad to correspond with the Secretarys of Chrysanthemum shows all through the South. We may have something of interest to offer them.

Our Chrysanthemum Book.

Give us your order for the new work, "Chrysanthemum Culture for America." The price is cheap, and it should be read by all.

GERANIUMS.

THERE is hardly a plant which is more popular among all classes on the globe than what is generally known as the Horse Shoe, Zonale or Fish Geranium. The Geranium is found under many different circumstances; it helps to embellish the conservatories of the millionaires as well as the homes of the humble and industrious, but it loses nothing of its inherent beauty on that account. It wanders with the household furniture from place to place, and the good wife makes a special request to her husband not to forget her Geranium. The propagation of the Geranium is universally known. Every woman knows how to slip it or grow it from cuttings; it is also produced freely from seed. They stand the hot sun of the South better than any other class of plants. They produce more flowers and make a better display on whatever place they are grown than anything that could be grown on a similar space. We have revised our Geranium list, and now have nothing but the very best and most distinct varieties.

NOVELTIES OF 1892.

Price 10 Cents Each ; $1.00 per Dozen.

Buffalo Bill. Double; immense globular trusses, creamy white, marbled in soft, bright rose; very fine.

Copernic. Single; florets are perfectly round, large and of finest form; beautiful shade of rose carmine, with a white eye.

Comte de Blacas. Single; orange salmon, shading to white on edge of the petals; white eye; very fine.

Golden Dawn. Double; an attractive shade of dazzling golden orange; very fine.

Gripper Banks. Double; produces the largest truss of orange scarlet flowers of any double now in cultivation; very fine.

Jean d'Arc. Single; a very much improved Souv. de Mirande, being much darker and freer in bloom; floret and truss of very fine size.

James Vick. Double; salmon, with deep brown shadings.

Lafayette. Double; very deep carmine pink; compact and stock plant.

Mrs. J. M Gaar. Single; the finest single white bedder yet introduced; compact, semi-dwarf habit; florets pure white and of medium size; a very free blooming variety; trusses very large, foliage dark.

Madonna. Single; a very soft shade of pale pink; base of petals white, trusses very large and handsome.

Montesquieu. Double; a soft lavender pink, of even shade, with white markings; very large trusses, of perfectly formed florets.

Mrs. E. G. Hill. Single; Bruant type; elegant salmon tint, with veining of a deep rose; florets very large; the best salmon yet introduced.

Mrs. A. Blanc. The flowers are of the very largest size, round and perfect; an apricot red, with touches of lilac at the centre; a grand single variety for bedding out.

M. Poirier. Single; a soft, vinous rose; the prettiest Geranium in this color, with markings on the upper petals that are beautiful.

Mad. la Countess de Pot. Single; salmon flesh color, veined with rose; edge of petals white.

Mous. Can. Double; a very deep rose, shading to white at base of petals.

M. Jovis. Double; immense trusses; the florets of salmon pink on a cream ground.

M. Louis Fages. A beautiful compact grower; fine foliage nicely zoned; the trusses large, florets of extra size; semi-double; color, clear orange; very free in bloom.

M. V. Noulens. May be described as a scarlet Mirande; a magnificent variety of large size, and of the most brilliant combination of clear white, with scarlet border.

Ruy Blas. Very large double florets in medium sized trusses; large centre of fiery salmon, the edges a soft and rosy salmon.

Souv. de Mirande. Single; the flowers round, trusses large, upper petals pure white, with a delicate edge of soft, salmon rose; lower petals bright salmon rose, with light shadings and a white centre; very popular.

S. G. Cobb. Single; the three lower petals are bright pink, the upper petals pink, marked with white centres; very fine, large and open trusses; a superior pot plant.

Univers. Single; a very dark crimson scarlet, of an attractive shade; very large truss, of finely shaped florets; an elegant bedding variety; heavy, compact grower.

W. A. Chalfant. Single; dazzling scarlet; florets circular and slightly cupped, forming a magnificent truss; very heavy, Bruant-like foliage; a very good bedder.

DOUBLE FLOWERING VARIETIES.

Price 10 cents each. Our Selection of Varieties, by Express, Twenty for $1.00; by Mail, Sixteen for $1.00.

Asa Gray. A light salmon dwarf; very free flowering.

Amelia Baltet. The best double pure white.

Belle Neneienne. The plant is dwarf and floriferous, with trusses of large, full florets of a fresh and very attractive color.

Bruant. A grand sort; trusses and pipe of immense size, semi-double, the color beautiful, brilliant and sparkling vermilion.

Bridal Bouquet. Very beautiful double white flowers, producing freely.

Centaur. Carries the largest and most perfect truss of any of the pink doubles.

Candidissima. A large, full and finely formed flower, the most snowy whiteness, not changing to pink.

Glorie de France. Large round flowers; color salmon white.

Grand Chancellor Faidherbe. A new sort; very thick and double flowers, of a dark soft red, tinted with scarlet and heavily shaded with maroon.

La Vienne. Dwarf and short jointed; a creamy white; semi-double.

Le Negro. A rich maroon; the darkest variety we have seen.

La Fraicheur. Plant short jointed and of very free growth, freely producing large trusses of well formed flowers of a tender lilac rose; a new shade of color and quite distinct.

La Favourite. The best double white; a good grower and profuse bloomer.

Mat Sandorph. Semi-double salmon.

M. Jasaine. Double pink.

M. Press. Beautiful double salmon.

M. Carr. Double dark crimson.

Marvel. A dark crimson maroon; extra fine variety.

Mad. L. de Beuregard. Large double florets of a lively salmon, each petal distinctly bordered with white.

Mrs. E. G. Hill. Ground color a pale blush, overlaid with a delicate lavender shade.

Mrs. W. P. Simmons. The flowers and trusses large; deep salmon with deep bronze shadings; plant of free blooming habit.

Mon. Gelein Lowagie. The brightest orange shaded.

Mad. Thibaud. A beautiful rich rose shaded with carmine violet.

Peter Henderson. A very fine variety with bright orange scarlet flowers of a fine shape.

Peach Blossom. Semi-double; beautiful rosy peach color; a good bloomer.

Richard Brett. Peculiar orange color, the nearest approach to yellow; a good bedder.

S. S. Nutt. Rich crimson, dark trusses, massive and profuse.

SINGLE FLOWERED.

Alphonse Daundet. Fine rosy salmon; very attractive.

Are-en-Ciel. The trusses of this variety exceed in size nearly all the single flowered; color lake red, the upper petals marked with orange scarlet.

Colonel Holden. A very beautiful rosy crimson; a distinct color; a very free bloomer.

Gen. Grant. Scarlet; a good bedder.

Gardner Gardett. Single, of a purplish crimson.

Heteranthe. Light gold dust red; cup shaped flowers; extra.

James Viek. The flowers and trusses of a great color, a deep flesh, with dark bronze shadings; of free habit.

Julius Lartigue. Rose color; vigorous and free.

Kate Schultz. The finest salmon pink; beautiful.

Louis Ulbach. Large scarlet.

Leon Perrault. Finest scarlet.

Master Christine. Pink; a very good bedder.

Mary H. Foote. One of the most beautiful salmon colored, even shade; very fine.

Mrs. Moore. Pure white, with a beautiful ring of bright salmon around a small white eye; of a dwarf habit; free flowering; very desirable.

New Life. A sport from striped Vesuvius, having its bright scarlet flowers striped and flaked with salmon and white.

Neve. Plant vigorous and of splendid habit; very large trusses of the purest white.

Poet National. Round florets, which are nicely displayed; color of Baroness Rothschild Rose, deepening to a soft rosy peach.

Queen of the West. A bright orange scarlet; very large truss; very profuse bloomer; we know of no finer for planting out in beds.

Reflector. Very bright and handsome scarlet, with a large pure white eye; trusses large and freely produced.

Sam Sloan. Fine bedding variety; deep velvety crimson; large truss and very free flowering.

Wm. Cullen Bryant. The finest shaped single flowering Geranium; a soft rich pure scarlet.

White Swan. Large bold white flower; fine for bedding.

IVY LEAVED.

Bijou. Hybrid; double scarlet.
Dolly Varden. Gold and bronze.

German. Or Parlor Ivy.
Remarkable. Flowers rose and white.

GOLD, BRONZE AND SILVER LEAVED.

Bijou. Flowers dazzling scarlet; leaves bordering white.
Golden Harry Tricolor. Golden yellow; bronze zone.
Mountain of Snow. The flowers of a bright scarlet; leaves margined with white.

Marshal McMahon. Yellow ground; a bronze zone.
Happy Thought. Centre of the leaf a creamy yellow, with a broad margin of deep green.
Mad. Salleroi. New Silver; great acquisition in variegated Geraniums.

SCENTED.

Lemon. 10 cents each.
Nutmeg. 10 cents each.
Oak Leaf. 10 cents each.

Pennyroyal. 10 cents each.
Rose. 10 cents each.
Mrs. Taylor. Rose scented; large scarlet.

PELARGONIUMS.

A Fine Assortment, 25 Cents Each.

APPLE SCENTED GERANIUMS.

We have succeeded in raising a large stock of this popular scented Geranium. No collection of plants complete without it. There is no one plant more popular nor more widely known than the Apple Scented Geranium. We have a handsome lot and offer them at a reduced price. Nice strong plants from three-inch pots, 15 cents each; a few extra large plants, 25 cents each.

NEW SCARLET BEDDING GERANIUM.

LEON PARRAULT.

There is nothing more beautiful on a lawn than a bed of solid scarlet Geraniums, and for this purpose we cultivate this variety extensively, as it stands the sun well and blooms freely the whole Summer through. A bed of it when in full bloom can be seen a long way off. Price 50 cents per dozen; $4.00 per 100.

CARNATION PINKS.

CARNATION PINKS, next to Roses, are the most popular flowers grown. The young plants should be procured in April or May, and be sure they are young plants, no matter how insignificant they may look, for large plants are ones that have been bloomed all Winter, and are comparatively worthless. Carnations are quite hardy, and should be planted as early as possible, just as soon as the ground is in condition to work. The soil should be quite rich, well manured with thoroughly rotted manure, or, if not to be had, bone dust may be used to a good advantage. To have a beautiful bed of Pinks in the Fall, the plants should be set out about eight inches apart each way; as the plants grow, they should be "stopped," that is, when the shoots of growth become six inches long, they should have the points pinched out. The operation should be continued until the 1st of July.

when it must be discontinued if the flowers are wished in August. Price 10 cents each, fourteen for $1.00, purchaser's selection. Our selection, by mail, sixteen for $1.00; by express, eighteen for $1.00.

Buttercup. Magnificent yellow.

Ben Hur. Grace Wilder pink, of even a brighter shade; indescribably beautiful; elegant flowers on long stems, freely produced; one of the best of all the recent introductions.

Dawn. Flowers very fragrant; color a blending of pink from centre outward to pure white on margin.

Daybreak. A genuine novelty in color, being a very delicate shade of pink, and is admired by every one.

Ferdinand Mangold. Flowers large and perfectly formed; color a brilliant red, shaded maroon.

Golden Gate. This is the finest of all the yellow varieties; full and double as Buttercup, but is a brighter golden color; it is a very vigorous and healthy grower.

Grace Farden. Flowers of medium or large size and very freely produced.

Grace Wilder. One of the most beautiful colors, a soft shade of carmine pink; a dwarf but robust grower; very desirable.

Hinze's White. Good, strong, dwarf.

Indiana. Of immense size; very tinely splashed; carmine on a creamy white background.

J. J. Harrison. Flowers a pure satiny white, marked and shaded with a rosy pink.

John McCulloch. This is the most brilliant and the finest scarlet yet introduced.

Lady Emma. Intense scarlet; profuse bloomer.

Lizzie McGowan. An elegant and new white variety that is destined to become very popular, being of the purest white color, large, full and very attractive; very prolific and fragrant.

L. L. Lamborne. Flowers large and of the purest white; dwarf habit.

Louise Porsch. After the style of Buttercup, but habit slender and wiry; the flowers very beautiful on extra long stems.

Mrs. Cleveland. Most charming shade of pink.

Nelly Lewis. A novel variety and entirely distinct from all others; a most pleasing soft shade of pink; is a strong grower, a free bloomer, and exquisitely fragrant.

President Degraw. Exceedingly fine white.

Petunia. This resembles a double Petunia so much as to be most appropriately named; of a very beautiful lavender rose, mottled white; deeply fringed.

Portia. Most intense bright scarlet; the flowers are of small size, but of fine shape and long stemmed.

President Garfield. Very decidedly the best of all scarlet Carnations in all respects.

Philadelphia. White, heavily edged and striped dark crimson.

Quaker City. A magnificent hardy white; very profitable for Spring forcing.

Sunrise. Orange, flaked with crimson; a new variety.

Silver Spray. White; the flowers are large, long stemmed and freely produced.

Starlight. A sport from Hinze's White; of the same free blooming and strong growing character; the shade is a clear even straw color.

Tidal Wave. A very dwarf variety, of a carmine pink color; very productive and early.

DAHLIAS.

Price 15 cents each; $1.50 per Dozen. Strong green plants, ready April 1st, 10 cents each; $1.00 per Dozen. The tuber or roots cannot be sent by mail.

Abbe Bertin. Maroon, very dark, free and early bloomer, dwarf plant about 2 feet high; flower ball-shaped with a long stem.

A. D. Livoni. While not exactly new it is the most lovely form of any pink Dahlia; has long stems, petals beautifully quilled, regular and double to the centre; an early and very profuse bloomer.

Bonnard's Yellow. Good yellow; it grows about 4 feet high; a desirable variety.

Camelliaflora. The plants grow to a uniform height of about two and a half feet, and are literally covered with its pure snow-white flowers during the entire season.

Constance. White Cactus Dahlia; a beautiful and desirable variety.

George Rawlings. Very dark maroon; full size; good all around bloom.

Gem. Bright scarlet pompone; an early variety of compact habit and very free blooming qualities.

Garibaldi. Yellow; dwarf, and blooms of good size.

Guiding Star. Pure snowy white, the petals toothed and fringed; beautiful and distinct.

Kate Haslam. Pink; medium sized blooms, freely produced.

King of Dwarfs. Rich deep purple; a very dwarf and free bloomer.

King of Cactus. Flowers bright crimson of the most intense color, and very large; petals very broad, somewhat twisted; stems long, superb for cutting; free bloomer; one of the very finest introduced for many years, and rightly named.

Lucy Fawcett. The most floriferous of all Dahlias; large and perfect; a light straw yellow flaked and streaked carmine.

Miss Ruth. Pure yellow, each petal tipped with white; early, free blooming and very pretty.

Mary Eustace. Fine straw color, tipped with white; tall and free blooming.

Princess Mathilda. Pure white of a dwarf habit; handsome foliage, and a good bloomer.

Purple Queen. Rich purple, occasionally shaded lilac; grows about 3 feet high.

Queen Victoria. Bright yellow; makes a handsome plant.

Red Head. Small red blooms, freely produced; plant rather a tall grower.

Snow Dwarf. A capital sort of dwarf habit from Germany; has long stem, very large pure white flower, smooth rounded petals.

Scarlet Giant. Fine rich scarlet; rather tall grower.

Victory. Scarlet; free bloomer; medium size blooms.

White Dove. Flowers are extra fine, double, pure white, petals beautifully toothed, stems very long; the cut flowers of this variety are more sought for by the florists than any other.

Wacht am Rhein. Tall growing variety; medium blooms; a rich purple in color.

VERBENAS.

W E have a large stock of these useful and popular bedding plants, and grow them extensively. The following comprise the best and most distinct colors of the new Mammoth strain, the distinguishing peculiarity of which is that the flowers are very much larger than the ordinary type, each individual floret being of the size of a silver quarter dollar, and the truss fully nine inches in circumference; they are of all the shades known to Verbenas. Price 10 cents each, sixteen for $1.00, purchaser's selection. Our selection, by mail, eighteen for $1.00; by express, twenty for $1.00.

Auricula. Fine large purple.

Admiration. A rich clear vermilion; large white eye; extra.

Blue Bonnet. Rich deep blue.

Beauty of Oxford. Dark pink, of immense size.

Bernica. Crimson maroon; very good flower.

Blue Bird. Blush purple.

Candidissima. Finest white.

Columbia. White, striped purple.

Century. Rich dazzling scarlet.

Crystal. Pure white.

Damson. Rich purple mauve, with a clear white eye.

Daisy Dale. Beautiful pink.

Endymion. Deep vermilion, crimson shaded, large white eye; extra.

Fanny. Violet rose, large white eye.

Glow Worm. Brilliant scarlet; perfect form.

Maltese. Lilac, shaded blue.

Mrs. Massey. Salmon pink, large white centre.

Marion. Mauve, of perfect form, white centre.

Miss Woodruff. Dazzling scarlet, very fine.

May Queen. Soft magenta pink.

Niobe. Deep vermilion; fine flower.

Nelly Park. Orange scarlet; splendid.

Purple Queen. Royal purple, with a large white eye.

Perfection. A rich chocolate maroon, lemon eye.

Rosy Morn. Pink, with a large white eye.

Scarlet King. Fine, vivid scarlet; dark eye.

Surprise. Clear, orange scarlet; white eye.

Sylphe. The best white Verbena in cultivation.

Snow Flake. Pure white; large truss; a fine and healthy grower.

CLEMATIS.

O F all the hardy running vines in cultivation, none is more beautiful than the Clematis, being entirely hardy and growing as they do more beautiful each year after being planted. They should be grown extensively. To anybody that has a position where a vine can grow, by all means, we say, plant a Clematis, for they are truly not only "things of beauty, but a joy forever." Large, strong plants, 50 cents each.

Azure. Light blue.

Aurora. Double red, shaded mauve.

Alexander. Pale reddish violet; a free bloomer.

Duchess of Edinburg. Double white.

Duke of Teck. White and mauve.

Gipsy Queen. Rich velvety purple.

Fair Rosamond. Blush white, red bar.

Gem. Deep lavender blue.

Hybrida Splendida. Reddish violet.

Helena. Pure white, colored anthers.

Henryii. Commences to bloom in June; its first bloom is immense, after which it blooms at intervals during the whole season; rich creamy white.

Jackmanii. Intense violet purple.

Jackmanii Alba. A lovely white, same as Jackmanii except in color.

Jeanne d'Arc. Grayish white.

Lady Londesborough. Gray, with pale bar.

Lady Caroline Neville. French white.

Lanuginosa. Pale lavender.

Lord Londesborough. Mauve, with a red bar.

Lucie Lemoine. White, with yellow anthers.

Mad. Granger. Purplish red.

Mad. Torreana. Bright rose.

Miss Bateman. Pure white, chocolate anthers; grand; this sort commences to bloom as soon as its leaves start in May, on which account it is very valuable.

Prince of Wales. Deep bluish mauve, with satiny surface.

Rubella. Rich scarlet purple.

Rubra Violacea. Maroon purple.

Star of India. Reddish violet, with red bars.

Standishi. Light mauve.

The Queen. Mauve, Linuginosa-like.

CAMELLIA JAPONICA.

T HE rich and pleasing contrast afforded by their dark green leaves and their superb flowers of exquisite beauty and waxy texture, together with their almost endless variegations of color, combine to make them one of the most desirable of Winter flowering plants. Price $1.00 each, nice bushy plants about sixteen inches high. We have no large size Camellias to offer this Spring.

Augustina Superba. Of a transparent rose color, sometimes spotted white.

Alba Plena. Large flower; white imbricated.

Archduchess Augusta. Beautiful red, with a dark azure vein and a white band in the middle of each petal, the flower assuming a blush and variegated color.

Angelo Cocchi. White, sometimes spotted or striped a bright red, sometimes dark.

Archduchess Marie. A magnificent flower of good form, very double; vivid red, with white rib-bands.

Auguste Delfosse. A fiery rose color, centre of petals striped; finely imbricated.

Aspasia. Small petals; very compact; brilliant red, rosy white heart.

Bonomiana. Large petals, rounded and imbricated in regular form; white line crossed through and through with deep red.

Chandleri. Flowers large and petals broad, of a rich pink color.

Comtesse of Orkney. Pure white with carmine stripes, often Peony form in the centre; very large petals, sometimes a very bright red, shaded dark red or white rosy stripes, with edges pure white.

Commendatore Betti. A superb variety, finely imbricated; red, changing to rose.

Candissima. Pure white; imbricated.

Comtesse Lavinia Maggi. Large buds, the flowers well formed, dotted cherry red.

Duchess de Berry. The flowers large, pure white; habit good, with fine foliage.

Fannie Bollis. A magnificent flower, well formed and large pointed petals of a flesh white, stained blood red.

Imperatrice Maria Theresa. Large and splendid imbricated flower; petals bright red, changing to pure white.

Imbricata. Carmine red, sometimes of variegated color, hence called Imbricata Tricolor.

Jubile. An extra large flower, imbricated; petals large and rounded; centre white; lightly rose sprinkled.

Leon Leguay. Very double; red, shaded deep red, exterior petals undulated; a first-class variety.

Leopold I. A bright scarlet red, with plush crimson bars near the border of the petals; extra fine variety.

Lyanna Superba. A vivid red; flower imbricated.

Mistress Cope. White flower, crimson stripe; of splendid form; extra.

Mad. Leboise. Imbricated; very bright red.

Mathotiana Alba. Pure white.

Noblissima. Pure white; Paeony form; very highly valued on account of its early flowering.

Princess Baciochi. A superb flower, well imbricated; a cherry red, with small white bands.

Prince Albert. Blush white, with numerous stripes of deep rose.

Princess Clothine. Imbricated, very nearly double; it has very strong petals with large white bands and deep red bars.

Reine Marie Henrietta. Of very fine form; of splendid foliage; a rose color, often speckled pure white; perfectly imbricated; very free bloomer.

Reine des Fleurs. Small leaves, but a vigorous grower, of good habit; a deep rich crimson color.

Trionfa di Lodi. Imbricated; large white petals, speckled and striped.

Union. Very large pure white flowers; sometimes Paeony formed; a first-class variety.

AZALEAS.

A ZALEAS are a class of plants highly ornamental for Winter and Early Spring flowering. They are of easy culture and can be had in bloom from Christmas to May if a fair selection of varieties is kept up. Our Azaleas will be this season, as usual, the very best for shape, variety and bud. Nice shaped plants, 50 cents each; large plants, $1.00 each.

A. Borsig. A very fine pure white and double variety.

Alba Illustrata. Flower of the purest white, occasionally sprinkled with a lilac rose.

Alba Illustrata Plena. Pure white; double; fine for forcing.

Baronne de Vriere. Flowers are enormous, snow white, petals very large, with undulated margin.

Bernard Andrea. Rosy purple, double; very beautiful.

Bernard Andrea Alba. Superb white flowers, very double; a most desirable and beautiful variety.

Charles Enke. Rosy salmon, marginated with white; very fine.

Charmer. Bright amaranth, the upper petals beautifully blotched with much deeper shade.

Comte de Chambord. A salmony rose color, striped and edged by a wide festoon of the purest white.

Dr. Moore. Intense rose, with a white and violet reflection.

Deutche Perle. A double pure white; a very free flowering and early blooming variety.

Eugene Mazel. Rosy salmon, upper lobes violet.

Flag of Truce. Pure white, double and very full.

Iveryana. White, with red stripes.

Jean Vervaena. Deep, rich crimson, edged with white; dark spot on upper petals.

Le Flambeau. Glowing crimson, bright and effective.

Mad. Iris Lefebvre. The flowers are extremely double, of a dark orange red.

Mad. Vander Cruyssen. Soft, glossy rose, tinted with amaranth.

Model. Bright rose.

Mad. Dom. Vervaene. Vivid salmon rose, white margin.

M'lle Marie Lrfebvre. A large flower of exquisite form and substance; pure white.

Marquis of Lorne. The flowers are of a beautiful orange, with saffron yellow blotch; the petals are very large and round.

Oswald de Kerchove. A very beautiful variety of lake rose with a fiery blotch; large and well formed.

Princess Charlotte. Very large flower of a beautiful rose; fine form.

Raphael. Alba Illustrata Plena. Pure white, double; fine for forcing.

Roi Leopold. Rich glossy crimson; of very fine form.

Superba. Bright rosy carmine, of good form; late.

Sigismund Rucker. The flowers are a lilac rose, strongly netted, bordered with white.

Vittata Crispiflora. White, shaded a purple and crimson.

W. Wilson Saunders. Very fine white variety, striped and blotched a vivid red.

FUCHSIAS.

THESE, when in full bloom, are the most graceful of all cultivated plants; nothing can surpass the beauty of well grown specimens. They delight in a light, rich soil, and may be grown either as pot plants or in a sheltered border. In either case they should be protected from the hot mid-day sun and from heavy currents of air. They require plenty of water and partial shade. We have greatly reduced our list of these plants, and offer now only what is good. Every variety here named is first-class, and the list contains a sufficient variety for all purposes. Price 10 cents each, 3 for 25 cents, 7 for 50 cents, 15 for $1.00.

DOUBLE.

Avalanche. Violet corolla, carmine tube and sepals, the foliage bright yellow.

Mad. Van der Strass. Tube and sepals deep scarlet; corolla nearly pure white; very large.

Mrs. E. G. Hill. Sepals dark red; corolla satiny white.

Purple Prince. Sepals red; corolla a dark purple.

Noveau Mastodonte. Sepals red, corolla purple.

Phenomenal. Short tube, and sepals of a bright reddish crimson; corolla very full, azure violet, flaked with red; very fine.

SINGLE.

Black Prince. Sepals carmine; corolla pink; bell-shaped.

Chas. Blanc. Light rosy pink sepals, corolla rich amaranth; a very robust grower.

Minnesota. Corolla light purple, sepals white.

Ernest Renan. Tube short, rosy white sepals, relieved; large rose-colored corolla.

Speciosa. Blush tube and sepals, light red corolla; very free.

W. E. Walt. Very rich purple corolla, sepals white, shaded rose.

NEW FRENCH CANNAS.

THERE has been nothing offered in recent years that possesses such real and intrinsic value and merit, and that appeals to so many buyers, as the new Cannas which we have the pleasure of offering to our customers this season. When the large size of the flowers and the large heads of bloom are taken in connection with the freedom with which it blooms, it makes one of the most attractive plants for bedding that it can be possible to conceive of. The other new varieties here offered are all equally desirable, and in presenting to our customers these varieties we are offering something which, while comparatively expensive, are well worth the cost, and we know no one will regret purchasing, even at the seemingly high price. They are really the finest novelties that have been offered in years; they are not only among the very finest plants for bedding, and will be in great demand by those who desire something choice and rare to vary the monotany of out door gardening, but they also make elegant plants for the conservatory, both Summer and Winter; in fact, one of the finest plants for conservatory decoration there is to-day. Every shoot blooms, and as often as a truss of bloom is past its prime, another shoot pushes out beyond it, and each stem blooms three, four, and five times in succession; as the plants throw up numbers of shoots, they are a mass of most gorgeous bloom constantly. We have never offered a plant that has given such unqualified satisfaction, and for which there has been such a demand as there has been for these new Cannas during the short time which we had them, and those who do not secure a number of these valuable plants immediately will have cause to regret it. Heretofore, we preferred to wait before investing in French Cannas until something was produced of intrinsic merit, pronounced enough in character to make it really valuable. Having satisfied ourselves of the value of these novelties beyond all question, we have invested in them very heavily, and we have now by far the largest stock of these in the country.

Childsii. Height four feet; the foliage rank, and of a light green shade; the flowers large, clear yellow, brilliantly spotted with crimson. Price 35 cents each.

Francois Crozy. This variety is identical with Md. Crozy in habit and general style of growth, but the flowers are bright orange, bordered with a narrow edge of gold, one of the most desirable shades that could be secured; it is similar to Md. Crozy in every way, except in color; many of the visitors to our establishment last Summer who saw it in bloom in the greenhouses, thought it was as valuable or more so than Md. Crozy on account of the desirability and rarity of the color in flowering plants, which is a very fashionable one; the plant is somewhat more dwarf than Md. Crozy, and was fairly covered with its orange-colored flowers, our tests showing it to be a more abundant bloomer than Md. Crozy, unless the brilliancy of the flowers seemed to make it so; it will be equally as indispensable as Md. Crozy. Price 60 cents each.

Gustave Senneholz. Green foliage; flowers salmon orange; large. Price 40 cents each.

Geoffrey St. Hillaire. This variety may be recognized by some, as it is not a novelty of this year, but has been offered for two or three seasons, but it is by far the best of the dark-leaved varieties among the new seedling Cannas that have been originated so far; notwithstanding it has been offered for several years, no particular attention has been called to it, and few are aware of its value; without taking into consideration the flowers of this variety, it would still be by far the best dark-leaved Canna, but when the large size of the flowers is taken into consideration, coupled with the richness of dark maroon-colored leaves with a dark metallic or bronze lustre, it will be appreciated by all lovers of rare and valuable plants; the flower is a light salmon scarlet. Price 50 cents each.

Henri L. de Vilmorin. An entirely distinct variety and of a most pleasing color; the centre of the flower is of a brownish red, shading off at the outer edge of the flower to a bright yellow, the two colors contrasting most beautifully; foliage pea green; height four feet. Price 50 cents each.

J. Thomayer. Flowers of very large size, produced in large spikes of a most intensely rich orange scarlet; foliage of a rich bronzy-purple; grows about four feet high. Price 50 cents each.

Louis de Merode. Leaves green, shaded with purple; large round flowers of amaranth crimson. Price 40 cents each.

Md. Crozy. The flowers are a flaming scarlet, bordered with gold, a marvelous combination of colors, having all the delicacy and beautifulness of the rarest Orchid; as a rule, the more valuable the variety, and the nearer it reaches perfection, the less vigorous it is, but in this case we have one of the strongest and most vigorous growing plants that has been produced in the Canna line so far; this Canna is one that every one can succeed with, as it will grow in the open ground as readily and easily as the common varieties of Cannas; these plants produce all the wealth and richness, and the tropical luxuriance of foliage of the common kinds, added to which is the gorgeousness of their flaming panicles of bloom, which are borne in immense heads at the terminus of every shoot. Price 60 cents each.

Maurice Rivoire. Rich bronze foliage, with large carmine flowers; it grows about four feet high. Price 50 cents each.

M. Duterail. Grows about two and a half feet high; compact deep green foliage; very vigorous grower; flowers large, of good form, deep saffron with a light margin. Price 50 cents each.

Mr. Lefebvre. Flowers of large size, of rich carmine; foliage rich bronzy purple; height four and a half feet. Price 35 cents each.

M'lle de Crillon. Green lanceolate leaves; large pale yellow flowers, spotted and shaded, red at centre. Price 35 cents each.

President Carnot. Deep purple foliage, with large carmine flowers; grows about four feet high. Price 40 cents each.

Professeur Chargueraud. Lanceolate green leaves; flowers large, of a vivid crimson; height three feet. Price 35 cents each.

President Cleveland. Height two and a half feet; green foliage; very largest salmon red flowers; an elegant sort. Price 35 cents each.

Perfection. Green foliage: the flowers large, golden yellow, minutely spotted red. Price 40 cents each.

Star of 1891. This variety was sent out last season for the first time, and was described in such a way as to seem identical with Md. Crozy, but it is entirely distinct from that variety; at the same time it has many valuable features of its own, distinct and characteristic; the individual flowers are nothing like as perfect as Md. Crozy, and they do not have the distinct golden edge that Md. Crozy has, although it shows a faint margin of yellow at times; the character of the heads of bloom are entirely different from Md. Crozy, growing first erect and then drooping; it is also more dwarf than Md. Crozy; it is nearly covered with a mass of scarlet, and as it is nothing like as tall a grower, when planted together will produce fine effects: it is an elegant pot plant and will bloom in the house in Winter splendidly, for which purpose it is admirably suited. Price 50 cents each.

Senator Millard. Fine purple foliage; a most ornamental variety, and free bloomer. Price 35 cents each.

The Water Lily Basin.

DO YOU GROW FISH?
IF SO
PLANT WATER LILIES.
They are the best known auxiliaries in fish culture.

DO YOU HAVE CHILLS?
IF SO
PLANT WATER LILIES.
They will absorb all the malarial poisons in the community.

DO YOU LOVE FLOWERS?
IF SO
PLANT WATER LILIES.
They are the most beautiful—the easiest cared for of all flowers.

CULTURAL NOTES.

ALTHOUGH Water Lilies may be cultivated in tubs, they may be grown to much greater perfection if allowed plenty of room, especially the larger growing tropical species. Those who wish to cultivate a number of kinds and have complete success, should build a tank about twenty by thirty feet, and two feet deep. If sunk entirely in the ground it would be more easily protected from frost in cold climates. But it may be partially sunken, and the soil which is taken out used as an embankment around the outside, sloping it up to the top. I prefer that it should be sunk to the level of the surrounding surface, for the reason that the banks can be made more ornamental. It may be built of either brick or stone. The bottom may be laid with rough stone, or old brick bats, and grouted with cement. Or, if the soil is of a firm nature, a thick coat of cement alone may be spread upon it. This latter plan has been perfectly successful with us, though we consider a concrete bottom preferable. Provide means for emptying the tank of water, when desired; also, a waste pipe, near the top, for overflow. After the walls have been built, and the bottom laid and grouted, the whole must receive an additional coat of cement. About four feet from each end of the tank, build a partition wall about twelve inches high. Bricks laid on edge will do, if laid in cement. These spaces can be cut in two by another partition. The compartments thus formed are for the purpose of confining the roots of the different kinds of Lotus and Nymphæa within proper limits, and for planting those kinds of Nymphæa which do better in such a position. The remaining portion of the tank can be taken up with pots, and large, shallow boxes, which will be movable at will. After the cement has properly hardened, fill the compartments and boxes with soil, and cover with an inch or two of clean sand. Fill the tank with water, and let it get well warmed before planting anything tender. As warm weather approaches, run a stream of fresh water in, for an hour or two each day, to prevent stagnation. When the surface of the water is covered with leaves, there is less tendency in this direction, and all that seems to be necessary is to replace what is lost by evaporation. The Lily tank must be placed in a warm and sunny position, for these plants will not do their best unless the water is thoroughly warmed. On the north side may be a border filled with Musas, Cannas, Bamboos, Ornamental Grasses, Caladiums, etc., which form a fine back-ground for the Lilies, and give the whole a tropical appearance.

SOIL FOR GROWING AQUATIC PLANTS.

The best soil for growing all kinds of aquatic plants in gardens we have found to be good, rich loam, and the best decayed stable or cow manure, in equal quantities, with the addition of about one pound of bone meal to a wheelbarrow load of the compost. Leaf mould or fine black peat can no doubt also be used to advantage. Rich mud from the bed of a pond or sluggish stream will answer in place of the loam, but we do not consider it essential. The compost should be well mixed, placed in the tank, and covered with about an inch of good, clean sand to keep the manure from rising; then let in the water several days before putting in the plants. The soil for the Nelumbiums should be heavy loam, or heavy, greasy clay, well enriched, as for all other aquatics. They will not flourish in sand or sandy peat.

WATER LILIES IN TUBS AND CEMENNT BASINS.

A good degree of success may be obtained by planting them in large tubs or half barrels in the open air, either on the surface or sunk in the ground. They should be placed where they receive the full benefit of the sun for at least the greater portion of the day. If for the whole day, so much the better. Fill them about half full of the compost recommended for all aquatics. The large growing kinds would do better in large half hogsheads or tierces sawed in two. A very effective and inexpensive plan is to arrange the tubs in connection with a rockery, a large tub in the centre being placed somewhat higher than the rest and connected by pieces of rubber hose, so that the overflow from the large tub runs from one to the other, so changing the water in all. Oil barrels make good tubs. The space around the tubs is filled with good rich compost, held in place by large stones, in which foliage plants, such as Sedums, Caladiums, Palms, etc., are to be planted. The effect produced in this manner is really beautiful.

ENEMIES OF AQUATIC PLANTS.

The conditions which we recommend for successfully growing tropical aquatics (i. e., still, warm water, and rich compost) favor the growth of a low form of vegetable life called confervæ, or green scum, which becomes very unsightly and troublesome unless eradicated. As the result of several years' experience, we are quite positive that if an abundance of Gold Fish are kept in the tank or pond, there will be no trouble in this direction. Other kinds of fish which are vegetarian in habit, might, perhaps, answer as well, but the German Carp is not to be recommended for tanks kept solely for the choicer varieties of aquatics, on account of their propensity for rooting in the mud and feeding upon the fibrous roots which proceed from the rhizomes of the Lilies. Should it be determined to keep a few

German Carp in the Lily Garden, it will be necessary to place whole pieces of roofing slate or large pebbles on the soil around the crowns of the tender Nymphæas.

Nymphæa Rubra. This species is a native of India, with flowers of a brilliant rosy red, stamens scarlet; the flowers are somewhat smaller than Devoniensis, measuring from six to eight inches, and are a trifle more cup shaped; the leaves are of a rich brown, and when they fade turn to a gold and crimson color, like Autumn leaves. $1.00 each.

Nymphæa Alba Candidissima. This is a large flowered variety of the Water Lily of England; when naturalized in still water, with a very rich soil, it produces leaves thirteen inches wide and flowers six inches in diameter, with very broad petals of a pure white color. 75 cents each.

Pontederia Crassipes. The Water Hyacinth. A very showy aquatic, bearing very freely flowers of a delicate lilac rose in trusses like a Hyacinth. The individual blooms are two inches in diameter; it should be grown in about three inches of water, so that the ends of the roots can enter the soil. 25 cents each.

Nymphæa Zanzibarensis Azurea. We offer under this name strong flowering bulbs, raised from seed of the true Zanzibarensis, which they are like in every respect, except that the color of the flowers is a shade lighter, being of the richest deep azure blue, far surpassing Cœrulea or any other blue Lily except the true Zanzibarensis; they are of the largest size and freely produced the entire year if the proper temperature is maintained; no collection is complete without this variety. $1.50 each.

Nymphæa Tuberosa Superba. This is by odds the best Nymphæa for the masses, as being so cheap it is within the reach of all; this is a decided improvement on the old Tuberosa; the flowers are larger, more fragrant, and more abundantly produced; broad petals of a pure white color; it blooms until frost, and lasts longer as a cut flower, than any other Water Lily; it is also good in fish culture, and as an absorbant of malaria; it may be grown in tubs, cement basins or natural ponds. 40 cents each.

Cyperus Alternifolius. Another well known greenhouse plant, which is really an aquatic or bog grass; planted out near the margin of the tank, or in pots submerged in the tank, it makes large growth, and is very ornamental and useful for cutting. 25 cents each.

Nymphæa Zanzibariensis. It is unquestionably the deepest colored and finest of all blue Water Lilies known, and some European horticulturists declare it to be the finest of the whole family; it is of a shade of blue so deep that it is not unreasonably called purple; it has the same fragrance as Cœrulea, and even when grown in small tubs or pans, produces larger flowers than that variety; under the treatment given it in our Water Lily garden they attain a diameter of twelve inches, and the leaves a diameter of two feet. $2.50 each.

Pontederia Crassipes Major. The Orchid Pond Lily. This is an interesting and beautiful aquatic; naturally a floating plant, the leaves having curious swollen stems, forming bulbs at the base, which are filled with air cells; the flowers are produced on very large spikes, like a Hyacinth, but are much larger; each flower is two inches across, and very chaste and delicate, color of a soft rosy lilac; often mistaken for an Orchid; it flowers best in shallow water, where the roots can take hold of the soil; will also do well in the greenhouse, or as a window plant. 25 cents each.

Nelumbium Luteum. The American Lotus; though a native of this country it is not common; there is scarcely any difference between this and Speciosum except in the color of the flowers, which are of a rich sulphur yellow; they are as large as a quart bowl, and have a strong fragrance, entirely unlike that of a Nymphæa; still, warm water and very rich soil are the conditions for success with these truly noble plants; a large patch of them, with hundreds of flowers and buds, is a sight never to be forgotten. Strong tubers, $1.00 each.

Nymphæa Cœrulea. This species is a native of Egypt; it has bright green leaves and lavender blue flowers, about four inches in diameter; they are very fragrant, the perfume being entirely distinct from Odorata; it is easily cultivated in a tub or tank, or it may be planted in a pond, where the water is still and warm and the mud rich. $1.50 each.

Nymphæa Devoniensis. This is the choicest Water Lily in cultivation; under liberal treatment it produces flowers ten to twelve inches in diameter, and leaves two feet across, the plant having a spread of twenty feet; if confined in pans, tubs, or boxes, the flowers are smaller, but otherwise just as fine; the color is a brilliant red, glowing by lamplight with indescribable brightness. $1.50 each.

Nymphæa Odorata Rosea. This is the famous Water Lily of Cape Cod; it possesses all the desirable qualities of Odorata; the flowers are of large size, cup-shaped; a lovely deep pink in color; of delicious fragrance and a free bloomer; a most desirable variety. Strong flowering roots. $1.00 each.

Nymphæa Odorara. Our native Water Lily; flowers pure white and very fragrant; when grown in rich mud it will produce flowers six inches and leaves thirteen inches across. 30 cents each.

Nelumbium Speciosum. This is the true Lotus of Egypt, and is one of the loveliest of the Water Lilies; the flowers when just expanding are of a beautiful rosy pink color; when open they are of a creamy white and pink, and are very fragrant; they open in the morning and close in the afternoon; it is a very interesting plant, and has been an immense success everywhere. Strong tubers, $1.00 each.

Special Low-Priced Collection of Aquatics.

The varieties named below are a selection of the most popular and easily grown sorts, which we grow in large quantities. We will supply the collection of six varieties, one plant of each, for $3.25, amounting at our regular price to $4.40.

Nelumbium Speciosum.

Nymphœa Tuberosa Superba.

Pontedeira Crassipes Major.

Nymphœa Odorata Rosea.

Nymphœa Cœrulea.

Cyperus Alternifolius.

LILIES.

L ILIES have long been celebrated for their chaste and rare beauty. It always has been and always will be a favorite. No plants capable of being cultivated out of doors possesses so many charms, rich and varied in color, stately and handsome in habit, profuse in variety, and of delicious fragrance, they stand prominently out from all other hardy plants, and no herbaceous border, however select, should be without a few of its best sorts. During the months of February and March, we can send by express Lilium Harrissii, grown in pots, with stems from one to two feet high, fine and healthy plants, for 50 cents each, that can be had in bloom at any desired time, according to the size of the plants selected.

Lilium Auratum. Gold banded; the finest of all. 25 cents each; $2.50 per dozen.

Lilian Candidum. The White Lily. 15 cents each; $1.50 per dozen.

Lilium Longiflorum. Pure white. 25 cents each; $2.50 per dozen.

Lilium Lancifolium Roseum. Blotched white and rose. 25 cents each; $2.50 per dozen.

Lilium Harrissii. The Bermuda Easter Lily. 25 cents each; $2.50 per dozen.

Calla Lily. Strong plants, 25 to 50 cents each.

Lily of the Valley. Strong clumps. 25 cents each.

Spotted Leaf Calla. (Richardia Alba Maculata.) This Calla resembles the White Calla, but is of a somewhat smaller habit; the leaves are of a glossy deep green, with numerous white spots, making it very ornamental at all times, even when not in bloom; it makes a splendid plant for borders or beds. 25 cents each.

Dwarf Everblooming Calla It is our good fortune to have come into possession of this most valuable of all Callas. The following facts will convince any one of its great superiority over the old sort. It is of a dwarf habit and does not grow tall and scraggy like the old variety, but is strong and compact, with a great abundance of lustrous dark green foliage. It is a true everbloomer, its flowers appearing in great abundance both Summer and Winter when grown in pots, or it will bloom profusely all Summer long in the open ground, and being potted in September will continue blooming without intermission all Winter The same plant will grow and bloom for years without once ceasing, and the quantity of bloom which a good plant will produce is astonishing. It is estimated that six plants of the old sort will not produce so many flowers in the course of a year as will one plant of this new dwarf variety. A large plant is hardly ever without one or more flowers, and its dwarf, compact habit makes it a much more desirable pot plant than the old variety. Its flowers are of large size and snowy white in color. All in all it is one of the most desirable plants we ever offered. 80 cents each; extra large. $1.00.

The Black Calla. (Arum Sanctum.) These magnificent flowers are sweet-scented, and often measure more than a foot in length by five to eight inches in width; the inside color is of the richest velvety purplish black imaginable, while the outside is of a pleasing green; the centre spadix rises to a height of ten to twelve inches, and is of the deepest black; the foliage is solid in texture, rich deep green, and elegantly reticulated. Strong bulbs, such as we send out, are positively sure to bloom if planted in very rich soil, two inches below the surface, and given plenty of light, heat and moisture; as the bulbs increase in size and number every year, such a treasure is sure to become valuable and appreciated. We price them as low as consistent with their cost, viz. 50 cents each; three for $1.00.

HELIOTROPES.

NEW AND SCARCE VARIETIES.

We offer this season the finest assortment of Heliotropes to be found in the country. Many of them were novelties of last year that proved to be of great merit. This is undoubtedly the best list of varieties to be found anywhere. We have propagated them extensively and are able to offer them at a very low price. Nobody should be without a few of these handsome new varieties. Price 10 cents each, fourteen for $1.00.

Albert Delaux. Bright golden yellow foliage, marked with delicate green, and lavender flowers.

Comtesse de Mortemarte. Very free flowering and exceedingly sweet; dark velvety blue, with white eye; splendid for massing.

Chieftain. A rich shade of violet; the best Winter bloomer.

Jersey Beauty. Finest blue variety; best for pot culture; dwarf.

Le Geant. The largest Heliotrope of which we have any knowledge; both the floret and truss are immense; it is of a lovely rosy violet shade of great beauty.

Souvenier. Half dwarf in habit, with panicles of immense size; florets large, centre a good white edged with bright lilac.

Mme. Ad. Dubouche. Foliage a very dark green, habit compact and shrubby; trusses enormous, soft and mossy looking; color velvety violet; one of the most beautiful Heliotropes ever introduced; color exquisite.

Mirielle. Panicles of the most extraordinary size and a very free grower; the floret is immense and quite flat, of a delicate pearl shade edged with Heliotrope; an improvement on White Lady though not so white.

Mrs. David Wood. Flowers are semi-double, in large heads, fragrant, curly, constant bloomer, light blue.

President Garfield. A gem of the very first water; fine deep blue, and very floriferous.

White Lady. Strong growing and free branching, and very profuse in bloom; large and of the purest white.

BEGONIAS.

This class of plants is each year becoming more deservedly popular. The beauty of their foliage and graceful flowers make them useful plants for greenhouse or window decoration.

Alba Picta. A perfectly distinct variety; leaves glossy green, thickly spotted with silvery white, the spots graduating in size from the centre towards the margin; flowers white. 15 cents each.

Cisse, Louise Erclody. Rex variety. This is the Begonia of all Begonias; its striking peculiarity, which distinguishes it from all other Begonias, consists in the two lobes not growing side by side, but one winds itself in a spiral way compotely over itself. 25 cents each.

Dregei. This variety is always in flower, Winter and Summer; it is one of the most useful plants we have; flowers white. 15 cents each.

Fuchsioides Rubra. Red flowers; very fine, and a constant bloomer. 10 cents each.

Fuchsioides Alba. Fuchsia like, pure white flowers. 10 cents each.

Glaucophylla Scandens. A very early flowering and vigorous growing variety, producing its beautiful clusters of salmon colored flowers from the axil of each leaf; its drooping habit makes it a very desirable plant for hanging baskets. 15 cents each.

Hybrida Multiflora. Flowers rose; it is in bloom almost constantly. 10 cents each.

Ingramii. One of the best Winter flowering varieties; flowers a reddish carmine, the leaves edged with bronze. 10 cents each.

Incarnata Metallica. Very dark green leaves, with silver dots and a metallic shade; fine pink flower clusters. 10 cents each.

Le Comte. Rex variety. The leaf is of medium size, of pointed Rex form; the color is very dark velvety green, just edged in very bright silver in the young leaves; the matured leaves are almost completely covered with bright silvery blotches on a dark background. 25 cents each.

Louise Closson. The brightest of all the rosy leaved Rex of which Chretien was the forerunner; then came Lucy Clossen, and now we have Louise Clossen, which has the Rex form of leaf nicely pointed; the texture is very full and crape like while the zone is a bright rosy purple of high metallic lustre. 25 cents each.

Metallica. A shrubby variety; a good grower and free bloomer; leaves triangular, longer than they are wide; the under side of the leaves and stems are hairy, the surface of lustrous metallic or bronze color, veined darker; flowers are white, covered with glandular red hairs; it is perfectly distinct. 15 cents each.

Nitida Alba. A strong grower and profuse blooming variety, producing immense panicles of pure white flowers; fragrant. 10 cents each.

Rubra. One of the finest Begonias in cultivation; its dark and glossy green leaves, combined with its free flowering habit, make it one of the very best plants for house or conservatory decoration; the flowers are of a scarlet rose color, and are produced in profusion 15 cents each.

Parvifolia. A dwarf, bushy growing variety, with pure white flowers, being in bloom the whole year. 10 cents each.

Perle Humfeld. One of the handsomest Begonias ever sent out; shows from six to eight deep points, arranged in very elegant form; the color is velvety green of various shades, broadly zoned with silvery spots. 25 cents each.

Ricinifolia. Has large palmated leaves, supported on stems from three to four feet long. 10 cents each.

Sutton's White Perfection. A beautiful dwarf, free flowering plant, that is always in bloom, and attracts attention wherever seen. 25 cents each.

Semperflorens Gigantea Rosea. Superb variety, of very strong, upright growth, fine large flowers of a clear cardinal red, the bud only exceeded in beauty by the open flower, which is borne on a strong, thick stem; leaves smooth and glossy, and attached to the main stem; both leaf and stem quite upright growing, forming a shrubby round plant. 25 cents each.

Saundersonii. Flowers a scarlet shade of crimson, borne in profusion during the entire Winter months. 10 cents each.

Sterling. A broad leaf, three pointed and the upper part lobed; a nice upright grower with smooth stems; the entire leaf is of a pale silvery green, of frosted lustre, broken only by pink ribs narrowly banded in green and narrowly edged red. 25 cents each.

Weltoniensis. Handsome Winter flowering variety; lovely pink flowers; of easy cultivation. 10 cents each.

Zebrina. Erect, of beautiful variegated foliage; leaves shaped like those of Rubra, and bearing white flowers. 10 cents each.

COLEUS.

C OLEUS can be used in numberless ways. Their foliage is superbly colored with bronze, crimson, maroon and gold, sometimes uniform, as in Verschaffeltii and Golden Bedder, but usually marked and variegated in the most brilliant manner. Beautiful at all seasons of the year, carrying rich and velvety foliage, and growing rapidly. They are very sensitive to frost, and should not be placed out in the open ground until all danger of cold weather is past. We can supply these plants by the thousand, and parties desiring to plant largely for Summer resorts or hotel grounds would do well to write for special prices. J. Goode is the best light colored Coleus for Summer bedding in the South, and Verschaffeltii the best dark; they both stand the sun well and make an excellent contrast. Price 5 cents each; 50 cents per dozen; $3.50 per hundred. It will be noticed that we offer these popular bedding plants cheaper than any other firm in the country.

Crimson Bedder. Rich dark crimson, with deeper shadings.

Corsair. A rich velvety crimson, bordered with deep maroon.

Firebrand. Bright glowing crimson.

Garland. Very large finely serrated foliage of bright green, purple and crimson.

Golden Bedder. Clear golden yellow; the best yellow bedder; a shy grower.

Hiawatha. Fine; orange yellow and crimson-flamed margin; this is one of the best and most distinct.

John Goode. Light green, on a yellow ground; a splendid bedder; a vigorous grower.

Kentish Fire. Centre clear vermilion, outside green and bronze.

Miss Retta Kirkpatrick. A creamy white, centre margined with green.

Novelty. Distinct green, covered with dots of yellow, maroon and red.

Onward. Very large foliage, of a deep green, marked and splashed with a brown.

Pluto. Green ground flaked and bordered bright carmine and chocolate; base yellowish white.

Princeps. Dark; bright crimson, yellow margin; one of the best.

Queen of the Lawn. Highly colored carmine, black and green.

Rob Roy. Bright carmine, edged with yellow and green.

Rag Carpet. Beautiful; carmine, white centre and green.

Spotted Gem. Yellow, blotched crimson, green and orange; very effective.

Verschaffeltii. Velvety crimson; the finest Coleus for all purposes.

PALMS AND DECORATIVE PLANTS.

THE following is a select list of rare and handsome varieties, which can be recommended for apartments, conservatory decoration, or vase plants during the Summer. All are in a clean and thrifty condition suitable for making immediate effects, and require no nursing to bring them into proper shape. The Seaforthias, Arecas, Latanias and Kentias are of quick and graceful growth, and can be grown without much trouble.

Areca Lutescens. One of the most beautiful and valuable Palms in cultivation; bright glossy green foliage and rich golden yellow stems. $2.00 each.

Areca Sapida. A strong upright growing variety with dark green feathered foliage. $2.00 each.

Aspidistra Lurida Variegata. Beautiful plant with large, lance shaped leaves, finely variegated with clear cream colored stripes; an elegant window or conservatory plant of the easiest culture. 50 cents to $1.00 each.

Aspidistra Lurida. A green leaved variety of the above, of strong growth; will succeed in any position; an excellent hall or corridor plant. 50 cents to $1.00 each.

Ardisia Crenulata. A very ornamental greenhouse plant, with dark evergreen foliage, producing clusters of brilliant red berries; a first-class house plant in Winter. 75 cents each.

Crotons. Of this beautiful class of ornamental foliage plants we offer twelve of the finest varieties. 30 to 50 cents each; $3.00 to $5.00 per dozen.

Chamaerops Excelsa. A handsome Fan Palm, of rapid, easy culture. $2.00 each.

Curculigo Recurvata. Very graceful Palm like plant for decorative purposes. 50 cents to $1.00 each.

Cycas Revoluta. The stem of this variety is very thick, and bears the foliage in whorls at the top. $5.00 each.

Caryota Urens. The Fish Tail Palm; an easily grown and useful sort. $1.00 to $2.00 each.

Dracæna Shepherdi. A strong growing variety, with deep green foliage, edged and striped with light green or yellow. 50 cents to $1.00.

Dracæna Terminalis. Rich crimson foliage marked with pink and white. 30 to 50 cents each.

Dracæna Fragrans. A superb African species, with very beautiful deep green leaves, lighter in the young growth; although having no variegations or markings in the foliage, this is one of the most admired of the decorative species, its elegant habit and extremely beautiful lively coloring giving it a marked value; it is a rapid grower, and for room adornment or as a vase plant for out of door use it is indispensable. 50 cents to $1.00 each.

Dracæna Youngi. Light green changing to copper color. 50 cents to $1.00 each.

Dracæna Australis. Very long, narrow, graceful foliage. 50 cents each.

Dracæna Indivisia. Long green foliage; graceful. 50 cents each.

Dracæna Veitchi. Long foliage, brown streaked. 50 cents each.

Ficus Elastica. The well known India Rubber Tree; one of the very best plants for table or parlor decoration; its thick leathery leaves enable it to stand excessive heat and dryness, while its deep glossy green color always presents a cheerful aspect. 50 cents to $1.00 each.

Ficus Repens. A trailing or creeping variety with small foliage; useful for baskets. 25 cents each.

Kentia Balmoreana. Beautiful strong growing Palm, with deep green crisp foliage. $3.00 to $5.00 each.

Kentia Fosteriana. One of the finest of Kentias, with graceful, bright green foliage. $3.00 each.

Phœnix Rupicola. Of graceful arching habit. $2.50 each.

Latania Borbonica. The Chinese Fan Palm; the most desirable for general cultivation, and especially adapted for centres of baskets, vases and jardinieres. $1.00, $2.00, $3.00 and $5.00 each.

Phœnix Reclinata. Beautiful reclinate foliage. $2.50 each.

Pandanus Utilis. Screw Pine. Called Screw Pine from the arrangement of the leaves on the stem. Excellent for the centre of vases and baskets, or to be grown as a single specimen; a very beautiful plant. $1.00 each.

Pandanus Javanicus Variegatus. Its leaves are green, with broad stripes of pure white, gracefully curved. $1.00 to $4.00 each.

Pandanus Veitchii. This is one of the most attractive plants; the leaves are light green, beautifully marked with broad stripes and bands of pure white, and very gracefully curved. $1.50 to $5.00 each.

Seaforthia Elegans. One of the very best for ordinary purposes; of graceful habit, and rapid, easy growth. $2.50 to $5.00 each.

GENERAL LIST OF PLANTS

—SUITABLE FOR—

Greenhouse or Out-Door Culture.

ALOCASIA.

Antiquorum. The foliage of this sort is of dark bluish-green color, grown to gigantic size; habit like the Caladium Esculentum. 50 cents each.

Odora. (Caladium Arboreum.) This noble plant while young slightly resembles the well known Caladium Esculentum, but grows to gigantic dimensions as it attains age, and while the latter dies down annually to the bulb, it is grown into a stem or trunk, which retains the foliage through the Winter if kept in the conservatory or eating room. The leaves are of enormous size, of a bright, glossy green, with thick fleshy midribs and nerves standing stiff and upright on a stout stem plants, under good culture, frequently attain eight and twelve feet in height 50 cents each.

ANTHERICUM VARIEGATUM.

Very valuable as a decorative plant, it being suitable either for the greenhouse, parlor or dining table. The foliage is of a bright grassy green, very beautifully striped and margined with a creamy white. 25 cents each.

ASPARAGUS TENUISSIMUS.

We cannot praise too highly this beautiful new plant. Its fine and filmy foliage equals in delicate beauty the Maiden Hair Ferns. 25 cents each; small plants, 10 cents each.

AGAPANTHUS UMBELLATUS.

A noble plant belonging to the bulbous rooted section, with evergreen foliage; the flower stalks grow nearly three feet high, crowned with a fine head of 20 or 30 blue flowers. 25 cents each.

ALYSSUM.

Double and very beautiful variety; is splendid for cut flowers; fine green foliage, and produces enormous quantities of double, pure white, fragrant flowers. 10 cents each.

ABUTILON.

Fairy Bells. Hard wooded greenhouse shrub, blooming almost the entire year; well adapted for house-culture, and fine for bedding out in the Summer. 15 cents each.

Boule de Neige. Pure white bell-shaped flower, blooming profusely.

Darwinii. Orange scarlet, pink veined flowers; blooms freely in clusters.

Mesopotanicum. Of a trailing habit; flowers pendant in great profusion.

Golden Fleece. A bright yellow; very profuse bloomer.

ACHYRANTHUS.

Suitable to form ribbon lines in contrast with Centaureas, Cineraria, Candidissima, etc. 50 cents per dozen; $4.00 per hundred.

Lindenii. Rich dark red color; is well adapted for either ribbon rows or the edging of flower beds.

AGERATUM.

Very useful plants for bedding or borders, flowering continually during the Summer; by cutting back and potting in the Fall they will continue to flower in Winter. 50 cents per dozen; $4.00 per hundred.

White Cap. By far the best and most useful variety ever sent out; a dwarf, compact grower, bearing profusions of pure white flowers.

John Douglas. Azure blue; of compact habit.

Meriden Gem. Compact; light blue.

ALTHERNANTHERA.

Plants with beautiful variegated foliage, growing from twelve to twenty-four inches in diameter and six inches high; it is used principally for ribbon lines and borders. 5 cents each; 50 cents per dozen; $4.00 per hundred.

Amabilis. Leaves tinted rose.

Aurea Nana. Foliage a bright green, beautifully variegated with yellow.

Paronychioides Major. Bronze, tipped red; the brightest and showiest.

ACHANIA.

Malvaviscus. Greenhouse shrub with scarlet flowers; it blooms Summer and Winter; not subject to insects of any kind; is one of the most satisfactory house plants ever grown. 25 cents each.

AMARYLLIS.

Johnsoni. An elegant pot plant, with crimson flowers five inches in diameter, each petal striped with white; the flower stalk is two feet high. 75 cents each.

Vittata. These magnificent flowers are flaked and striped with the most striking tints, and are justly esteemed. 75 cents each.

AGAVE.

This is better known as the Century Plant.

Americana. A picturesque plant for out-door decoration on the lawn, or for growing in vases. 25 cents to $1.00 each.

Americana Variegata. Similar to the above variety, with leaves banded with yellow; these plants stand any amount of heat and drouth, and are therefore admirably adapted for centre plant of vases, baskets, rock work, etc. Small plants, in four-inch pots, 35 to 50 cents each; large, one to two feet high, $1.00 to $5.00 each.

ASTERS.

Plants grown from the choicest seed, 50 cents per dozen.

ALOYSIA CITRIODORA.

Lemon Verbena. A favorite garden plant, with delightfully fragrant foliage; fine for bouquets. 10 cents each.

BLETIA TANKERVELLEA.

A beautiful terrestrial Orchid of free growth; the flower-stems are about eighteen inches high; the flowers are singularly beautiful, of a white and brown color, and bloom in the Spring. 50 cents each.

BONAPARTEA JUNCEA.

A very graceful genus of plants, with long, graceful, rush-like leaves; is very attractive when grown in a vase out of doors in the Summer; flowers borne in large spikes; it requires a warm temperature in Winter. 50 cents each.

BRUGMANSIA SUAVEOLENS.

A magnificent plant, growing four to six feet high; leaves large sea-green and velvety; flowers large, trumpet-shaped, double, and highly fragrant, about eight or ten inches long and five or six inches across the mouth; flowers pure white when fully expanded; profuse in Summer, and in a sunny window, a profuse Winter bloomer. 25 cents each.

BILBERGIA SPECIOSA.

Pineapple resembling foliage, with the very brightest crimson flowers growing out of the heart of the plant; of easy cultivation. 30 cents each.

BOUVARDIAS.

These are among the most important plants cultivated for Winter flowers, owing to the yearly increasing variety of color and excellent adaptation for that purpose. They are also effective as bedding plants for the garden, blooming from July until frost. 15 cents each; $1.50 per dozen; small mailing plants. $1.00 per dozen.

A. Neuner. Perfectly double; a pure waxy white; a constant bloomer, and of unsurpassing beauty.

Bockii. New single pink, producing its flowers in graceful clusters.

Lelantha. Dazzling scarlet; one of the best, and very profuse.

President Cleveland. Extra large fiery scarlet flowers; vigorous growth.

President Garfield. Double pink.

The Bride. White, with a very slight tinge of flesh; a really fine sort.

Vreelandii. Finest white; valuable for bouquets; best of all singles.

CALADIUMS.

Fancy Leaved. We have a fine collection of first-class, distinct. They are never as large as Esculentum, but the brilliant cardinal red, pink, cream and various shades of green that are displayed in the veinings and blotches of the leaves can not be obtained in any other class of plants. 30 cents each for fine, well dried tubers.

CALADIUM ESCULENTUM.

The most striking and distinct ornamental foliage plant in cultivation; is desirable for pot or tub culture, and fine for bedding out; with a plentiful supply of water, the leaves may be grown from four to six feet long, and one and one-half feet in breadth. 25 cents each; large bulbs, 40 cents each.

CYPRIPEDIUM INSIGNE.

Lady's Slipper. A terrestrial Orchid of easy cultivation. 50 cents each.

CACTUS.

Of these plants we have a fine collection. The Cactus family is interesting on account of the curious leafless growth of the plants and the beauty of the flowers, the Lobster Cactus, especially, being a great favorite.

Epiphyllum Truncatum. Lobster Cactus. Winter blooming. 25 cents each.

Cereus Grandiflorus. Night Blooming Cereus. 25 cents each.

CENTAUREA.

Gymnocarpa. Dusty Miller. Attains a diameter of two feet, forming a graceful round bush of silver grey, for which nothing is so well to contrast in ribbon lines with dark foliaged plants. 50 cents per dozen.

CESTRUM PARQUI.

The Night Blooming Jessamine. This is a well known and very highly prized plant, producing its richly fragrant flowers at every joint; sweet only at night. 10 cents each.

CINERARIA.

Hybrida. These are among the most gorgeous of our greenhouse plants; the colors range through all the shades of blue, violet crimson, pink, maroon and white. They are in bloom only until May. 15, 25 and 30 cents each.

CYCLAMEN PERSICUM.

As an ornamental greenhouse plant it is excelled by few, and its flowers as a variety in the formation of bouquets and baskets of cut flowers in Winter are valuable. 10 to 25 cents each.

CHRYS. FRUTESCENS.

This is the Paris Daisy now so fashionable and in such demand during the Winter. The flowers much resemble our common field Daisy; almost constant in bloom. 10 cents each.

COCCOLOBA.

Platyclada. Plant of singular and interesting growth; stem and branches growing to flat, broad points. It is well suited for vases and rustic work. 10 cents each.

CANNA.

The Canna is a fine foliage plant, making a good bed alone, but is particularly desirable as the centre of a group of foliage plants, for which it is one of the best, growing from three to six feet. Select old sorts, 10 cents each.

CUPHEA.

Platycentra. The Cigar Plant. Tube of scarlet, tip white and black; is very free blooming. Is a good basket plant and is also an excellent plant for the house in Winter. 10 cents each.

EUPHORBIAS.

Plants of great value for Winter blooming, and make splendid pot plants; they are sure to bloom with regularity, are easily cared for, and do not suffer much from a moderate amount of neglect or abuse. 25 cents each.

EUCHARIS.

Amazonica. A bulbous rooted plant, with very broad Lily-like leaves and pure white flowers about four inches in diameter, borne in heads of four or five, and deliciously fragrant. Fine bulbs, 50 cents each.

ECHEVERAS.

A genus of succulent plants, natives of Mexico. They are of rich appearance, and well suited for rock work.

Sanguinea. Narrow pointed leaves, of a deep red color. 15 cents each.

Secunda Glauca. Dwarf sort, resembling the house leek; a glaucus green; they bloom all Summer; an excellent plant for borders or rock work. 10 cents each; $1.00 per dozen; $6.00 per hundred.

FARFUGIUM GRANDE.

A highly decorative plant with round leaves, large as tea saucers, of a dark green color, profusely blotched with yellow; a great acquisition; of the easiest culture; fine either in or outdoors. 25 cents each.

FRAGARIA INDICA.

Indian Strawberries; of trailing habit; bears fruit throughout Summer and Autumn; very fine for baskets. 10 cents each.

FEVERFEW.

Little Gem. The finest double white raised; a first-class plant, that everyone should have. 10 cents each.

Double White. A very free blooming, double, Daisy-like flower; very useful for Summer bouquets. 10 cents each.

FORGET-ME-NOT.

Myosotis Palustris. Requires no description. Its clustered flowers of beautiful blue having had a place in romance and literature since romance and literature began. 10 cents each.

FERNS.

These beautiful plants are now very generally cultivated, their diversity of gracefulness of foliage making them of much value as plants for vases, baskets or rock work, or as specimen plants for parlor and conservatory. 15 cents each.

GREVILLEA ROBUSTA.

The Australian "Silk Oak." A splendid ferny-leaved tree, evergreen, and especially adapted as a shade tree; thousands are now being annually planted; also used by florists for decorating apartments, etc. A magnificent pot plant. 75 cents each.

GLADIOLI.

Among bulbous flowers the Gladiolus deserves first place in popular favor; our collection is very fine, and contains a good assortment of colors, red, pink, striped, and many shades of light colors. By express, 75 cents per dozen; by mail, $1.00 per dozen.

HYDRANGEAS.

Hortensia. The well known garden variety; has immense heads of pink flowers, which hang on for months. 15 cents each.

Otaksa. Heads large, bright rosy pink, contrasting beautifully with the other sorts; of a low bushy growth. 10 cents each.

Thomas Hogg. Immense truss, at first tinged with green, then turning a pure white. 15 cents each.

GLOXINIA.

Gloxinias are among the handsomest of our Summer blooming greenhouse plants. They require partial shade and a liberal supply of water when growing. After blooming, water should be withheld, and the bulbs remain dry during the Winter. 50 cents each.

HIBISCUS.

A beautiful class of greenhouse shrubs, with handsome glossy foliage, and large, showy flowers, often measuring over four inches in diameter; they succeed admirably bedded out during the Summer. 10 cents each.

Brilliantissima. Single flowers, of the richest crimson-scarlet; a dark crimson at the base of petals; very large and showy.

Denisonii Rosea. Large, single flower, of a clear transparent rose.

Grandiflora. Enormous rosy crimson, single flowers, which are produced in abundance.

Kermesinus. Enormous and very double rich carmine crimson.

Miniatus Semi-Plenus. An immense semi-double flower; a dark vermilion scarlet.

Zebrinus. Outer petals scarlet; edged yellow, variegated yellow and scarlet.

JASMINUM.

Grandiflorum. The Catalonian Jessamine. The flowers are pure white and most deliciously fragrant. 15, 25 and 50 cents each.

Grand Duke. Flowers double, white, like a miniature Rose, and deliciously fragrant. 75 cents each.

JESAMINE, CAPE.

Gardenia Florida. Southern plant of easy cultivation, blooming profusely in the Spring and early Summer: the flowers a pure white; double; plants bushy; foliage dark green and glossy. Plants that will bloom, 25 and 50 cents each.

Jasminum Revolutum. Beautiful yellow flowered hardy shrub, and a great favorite in the South.

LANTANAS.

Plants much used for bedding and pot culture. They are strong growing and constant bloomers. 10 cents each; $1.00 per dozen.

Aurantiaca. Beautiful crimson.

Jacob Schultz. Red, changing to crimson.

Purpurea. Good purple.

Rosa Mundi. White and rose.

MUSA ENSETE.

The noblest of all plants is this great Abyssinian Banana; the leaves are magnificent, long, broad and of a beautiful green, with a broad crimson midrib; the plant grows luxuriantly from eight to twelve feet high. We offer a fine lot of these plants at $1.00, $1.50, $2.00 and $3.00 each.

MUSA CAVENDISHII.

A dwarf variety with large ornamental foliage; can be cultivated in a tub or box, and will bear fruit; ornamental as a house plant in Winter. $1.00 and $2.00 each.

MARANTA ZEBRINA.

A house plant of unsurpassed beauty; foliage a dark velvety green, with black stripes. 50 cents each.

MESSEMBRYATHEUM.

Cordifolium Vagiegatum. A succulent plant; the leaves are distinctly variegated with green and white. 10 cents each.

NASTURTIUM.

Empress of India. The plant is of a very dwarf habit, with dark tinted foliage, while the flowers are of the most brilliant crimson, so free produced that no other annual in cultivation can approach it in effectiveness. 10 cents each. Double varieties, 10 cents each.

OXALIS.

These plants are of the easiest possible cultivation, and are fine for baskets, vases, etc.

Lutea. Large, clear, yellow flowers in the greatest profusion. 15 cents each.

Rubra. Flowers bright red.

White. White, flowers profusely Summer and Winter. 10 cents each.

OLEANDER.

Double Pink. The oldest and finest of all varieties in cultivation; the flowers are double, and rose colored. 20 cents each.

Lilian Henderson. A new double variety, and one of the best yet introduced. 50 cents each.

ORNITHOGALUM.

Star of Bethlehem. Remarkable for opening its umbels of satiny white flowers at 11 o'clock and closing them at 3 o'clock; blooms from May until July. 15 cents each.

PLATYCERIUM ALCICORNE.

The Stag Horn Fern of Australia. This most wonderful Fern has become so scarce that it is but seldom found in cultivation; the curious fronds and the strange habit of growth are really wonderful; while they also can be grown on blocks of wood, like Orchids, they are really in their element when grown in hanging baskets; when well-established, young fronds in shape of stag horns will appear everywhere, through the moss, making a very curious show; they are grand plants for house culture in a shady place; they require plenty of water. $1.00 each.

PILEA.

Arborea. The Artillery Plant. Pretty little plant of drooping habit, resembling the Fern; a fine basket plant. 15 cents each.

PLUMBAGO.

The Plumbagos are desirable on account of their beautiful shades of color, a color by no means too common among the flowering plants.

Capensis. Very bright plants with large heads of light blue flowers. 15 cents each.

Capensis Alba. In this new white sort we offer a sterling novelty which will heralded by all flower lovers with delight. 15 cents each.

POINSETTIA PULCHERIMA.

A new double Poinsettia; a very brilliant scarlet, tinted with orange color; a dazzling color; the head grows on a specimen plant 14 inches in diameter by 10 in depth, giving it the appearance of a cone of fire. 25 cents each.

PRIMROSE.

Chinese; few house plants afford more genuine satisfaction than this; it requires to be kept cool, a north window suiting it best; Primroses are at present all in bloom. 20 cents each; $2.00 per dozen.

PETUNIAS.

Double; the double flowers are of much greater size than the largest of the singles, and are very richly colored; they flower freely, and continue often even after hard frost. 15 cents each.

PANSIES.

This class of plants cannot be overestimated; the gigantic size of the flowers, its luxuriant growth, profusion of bloom, and exquisite blending of gay and fantastic color, is utterly indescribable; the colors are truly wonderful, including many different shades and combinations; we believe that our fine new Giant Pansies are the finest strain ever offered. 5 cents each; 25 for $1.00; $3.50 per hundred.

RUSSELIA JUNCEA.

Has long, very graceful, rush-like foliage, the drooping tips of which bear tubular, light scarlet blossoms in showers; there is nothing so beautiful for large vases, and a handsome house plant. 25 cents each.

RIVINIA LÆVIS.

A most charming plant, bearing long, pendent spikes of small pinkish white blooms, followed by brilliant red berries. 10 cents each.

RHNCOSPERMUM.

Jasminiodes. A greenhouse climber, with white Jessamine like flowers, which are produced in great clusters in the Spring months, and have a delicious fragrance. 25 cents each.

SALVIAS.

Flowering Sage. This plant is indispensable in the garden in Autumn; it may be planted in masses or scattered among the shrubbery; either way their gorgeous effect is well displayed. 10 cents each; where selection is left to us, 20 for $1.00.

Rutilans. Magenta; apple fragrance.

Splendens. Brilliant scarlet; beautiful.

Splendens Alba. White flowered.

TRADESCANTIA.

Zebrina. The Wandering Jew. The leaves are striped a silvery white. $1.00 per dozen.

VINCA.

Periwinkle; the best blooming plant for bedding out, being constantly in bloom from June until frost, bearing the hot sun and frequent drouth well, and is excellent for the South; we have a good stock. 10 cents each; 50 cents per dozen; $4.00 per hundred.

Alba. Pure white, hundreds on a plant.

Rosea Alba. Pure white, dark rose eye.

Rosea. Dark rose pink.

SMILAX.

A climbing plant, unsurpassed in the graceful beauty of its foliage; its peculiar wavy formation renders it one of the most valuable plants for bouquets, festoons and decorations. 15 cents each.

TUBEROSE.

Pearl. New double; the flowers of large size, imbricated like a Rose, of dwarf habit, growing only from 18 inches to two feet high. 10 cents each; 50 cents per dozen; $3.00 per hundred; by express, 30 for $1.00.

VIOLETS.

It is one of the leading florists' flowers for bouquets and cut flowers. All the varieties should have a slight protection of leaves during Winter; a better plan to insure early Spring flowers is, to plant in cold frames in the Fall; they thrive best in a shady situation, in rich deep soil. 10 cents each.

Blue Neapolitan. Double light lavender blue; very profuse bloomer.

Marie Louise. Double, darker than the above, and larger in size.

Schœnbrun. Single, dark blue, profuse.

Swanley White. Pure white, large size.

A Few Rare Trees and Shrubs.

ELEAGNUS LONGIPES.

A valuable new fruit, as well as a magnificent ornamental shrub, from Japan. Perfectly hardy, free from disease and insect vermin of all kinds. Very attractive, in bloom by May, after which, until late in Fall, it is clothed in luxuriant green foliage; silvery underside and producing in profusion handsome bright berries, which make delicious sauce. This fruit has been grown and highly prized in an amateur way for a number of years; but not until the past season has its great value as a garden or market fruit been recognized. We anticipate for this rare and valuable addition to our list of choice fruits an immense demand as soon as its great merits become known to the public. To some persons, even in its present state, the flavor is far preferable to that of the Currant or the Gooseberry. The plants are very productive, and they are easily raised and perfectly hardy. They possess, moreover, the merit of carrying their leaves bright and fresh well into the Winter. 75 cents each.

CITRUS TRIFOLIATA.

Hardy Orange. This extremely beautiful and curious Japanese Orange has proved perfectly hardy as far North as New York, Philadelphia, and Illinois, and may be seen growing in the parks in the former cities, and in the government grounds at Washington. It differs from other Oranges in having trifoliate, or clover-shaped leaves, larger and finer blooms than any other sort, and produced over a much longer season. The fruit is Orange red, about the size of a pigeon's egg, and of fine flavor; it makes a beautiful shrub about four feet high, completely covered with its sweet-scented blooms and brilliant fruits. It is very valuable for stocks on which to dwarf the larger growing varieties. Though hardy, it is best to protect it well over Winter, in the same manner as Rose bushes, etc. It can be grown as a pot or tub plant if desired, and made to bloom in Winter. At the remarkably low price at which we offer it everybody can own an Orange. Price of fine, thrifty plants, 50 cents each.

EXOCHORDA GRANDIFLORA.

This magnificent hardy shrub from North China, with its great racemes of snow white flowers, like single Roses, and its rich and long, persistent foliage and elegant habit of growth, ought to be in every garden. It blooms in May, and is one of the finest hardy shrubs of the present time. 50 cents each.

OLEA FRAGRANS.

This is best known as the Sweet Olive, and is a magnificent shrub that should find a place in every Southern garden. It is to well and favorably known to need description. Nice strong pot grown plants, twelve to fourteen inches high, 50 cents each; large plants, $1.00 each.

MAGNOLIA FUSCATI.

This is a plant that there is a wonderful demand for, owing to the delicious perfume of the flowers and rare beauty of the plants. This is the first time we have been able to offer it. Nice one year old pot grown plants, 50 cents each; large plants, 18 to 24 inches high, $1.00 each.

SOPHORA JAPONICA.

This tree has compound leaves of richest glossy green, and are as beautiful as anything in the whole range of foliage trees. The most unique characteristic of this little tree, however, is the color and smoothness of its twigs and branches, which remain for years a deep, shining green, and make the tree attractive even in Winter. Its blossoms are borne in long clusters of a rich, creamy color; for the lawn this really a perfect tree. $1.00 each.

POWLONIA IMPERIALIS.

A magnificent tropical looking tree from Japan; of extremely rapid growth, and surpassing all others in the size of its leaves, which are 12 to 14 inches in diameter; blossoms trumpet shaped, formed in large upright panicles, and appear in May, producing a beautiful effect. We have a few trees on our place planted out three years ago that are now 20 feet high and branched in proportion. The most rapid grower of all trees. $1.00 each.

AILANTHUS.

Tree of Heaven. A very handsome ornamental tree, having been introduced in this country from China many years ago, yet it is unknown in many parts. It is a very rapid grower, perfectly hardy in all parts, succeeding in all soils, and growing six to ten feet high with very stout stems the first Summer, with magnificent leaves five to six feet long, giving a good tropical appearance. In China it is known as the Tree of Heaven, a name suggested by its majestic form and great beauty. The trees grow to a good height and bear in the greatest profusion very large panicles of bloom. $1.00 each.

DOGWOOD—CORNUS.

The "Queen of Ornamental Trees" indeed. No other possesses so many virtues or is so nearly faultless. Thrice blessed is the Flowering Cornel, (even of perpetul though changing beauty) for in early Spring its galaxy of blossoms equals the finest Magnolia; in Autumn its foliage, almost dazzling in its brilliancy, surpasses the Scarlet Oak or Maple, while in Winter its clusters of bright vermilion berries add a charm and cheerfulness not otherwise to be obtained. To complete the circle of the year, in Summer its dense, handsome foliage affords always a perfect shade. Large enough in habit to be effective upon the largest lawn (specimens occasionally attaining a height of thirty feet) and yet small enough for grounds of the most limited dimensions, as by pruning it can be kept into almost a bush. It is hardy everywhere, from Canada to the Gulf, from the Atlantic to the Pacific, for no matter how hot or how cold it never yields to the vicissitudes of climate or weather. Flourishes upon all kinds of soil, and in all situations, wet or dry, upon hillside. among rocks, by streams or upon the level lawn; in rich loam, cold clay or poor sand. It is truly democratic, purely American. Red or white, strong and healthy plants, 75 cents each.

MAGNOLIA GRANDIFLORA.

The true Southern Magnolia of great beauty, too well known to need description in a Southern Catalogue. There is no home or door yard in the entire South but what should have at least one or two Magnolias. Plants that are sure to grow, from two to three feet high, 75 cents each.

CATTLEY GUAVAS.

These make fine pot plants and are perfectly hardy in the open ground anywhere where the thermometer does not go below 20 degrees. Plants evergreen, with beautiful shining, thick Camelia-like foliage, and in fruitfulness surpassing anything we ever saw. A plant eighteen months old has borne 500 fruits, and one a little older and larger held 1,005 fruits, blooms and buds at one time. After becoming established the plants will bear buds, flowers or fruit in some stage every day in the year. Strong plants. $1.00 each.

SPECIAL OFFER.

In order to give everybody an opportunity to enjoy the beauty of a few of these choice plants, we will send for ONE DOLLAR one strong plant each of Eleagnus Longipes, Citrus Trifoliate and Exochordia Grandiflora.

Hardy Herbaceous Plants and Shrubs.

Achillea, The Pearl. Grand improvement on the old Achillea: the flowers, which are borne in the greatest profusion the entire Summer on strong, erect stems, are of the purest white, somewhat resembling a Pompone Chrysanthemum; as a Summer cut bloom it is a great acquisition. 25 cents each.

Azalea Americana. Hardy Azaleas; deciduous, flowering in May; light straw colored blossoms; very beautiful. 50 cents each.

Astilbe Japonica. Incomparably the most beautiful of all hardy herbaceous plants, growing about two feet high, in compact shape, with handsome foliage, from above which rise its panicles of small and feathery white blossoms. 25 cents each.

Bridal Rose. Rubrus Grandiflorus. A large, double white flower of the Blackberry tribe; free growing; may be set out in Spring and potted off like Roses in the Fall; a good Winter bloomer 50 cents each

Crape Myrtle. Pink, fringed pink blossoms; Crimson, deep crimson. 10 cents each.

Calycanthus Floridus. Strawberry shrub; hardy, growing five to six feet high; dark brown flowers with a delicious odor. 25 to 50 cents each.

Deutzia Crenata. Height two to three feet, regular and compact form; bushy; flowers pure white; blooms profusely. 25 cents each.

Deutzia Gracilis. Very graceful white blooms, produced all the Spring in large quantities; dwarf and bushy. 25 cents each.

Dielytra Spectabilis. A hardy ornamental flowering plant; known also as the Bleeding Heart; a valuable garden plant. 20 cents each.

Erianthus Ravennæ. Perfectly hardy; the foliage forms graceful clumps three to four feet high, above which arise its numerous spikes five to six feet, bearing plumy flowers. 50 cents each.

Eulalia Japonica Zebrina. Unlike all other variegated plants, this has its striping or marking across the leaf instead of longitudinally; the extended flower spike resembles the ostrich plume, and will last for years. 25 cents each.

Gypsophilla Paniculata. Plants of the stitchwort family; forms a dense compact bush three feet or more in height and as much across; the flowers small, white, exceedingly numerous, arranged on thread-like stalks in branching stems, with the light, airy and graceful effect of certain ornamental grasses; is very useful for cutting. 35 cents each.

Helianthus Multiflorus. The Double Hardy Sunflower; an extremely useful hardy plant; it bears profusely large double flowers of a bright yellow. 15 cents each

Hydrangea Paniculata Grandiflora. One of the finest hardy shrubs in cultivation; flowers formed in large white panicles of trusses, nine inches to one foot in length. 25 to 50 cents each.

Holyhock. Superb double kinds; the Hollyhock is becoming a very popular Summer flowering plant, and when planted in rich soil and a sunny position is a very impressive and stately plant. We offer strong, one year old plants, at $1.00 per dozen; nice young plants that will bloom this year, 50 cents per dozen.

Iris. Fleur-de-Lis. The Iris is a very extensive and beautiful family, commonly known as the Flowering Flag. A beautiful assortment of color from pure white to richest purple. 25 cents each.

Phlox. Our collection embraces the best of the old varieties and the new French ones of recent introduction, which are very fine, distinct, and pure colors, many of them finely shaded and marked with distinct clear, light eyes. 15 cents each, $1.50 per dozen.

Lilac. Well known and popular hardy flowering shrubs. Large strong plants, 25 cents each.

Pæonias. Pæonias, like other meritorious plants, always have admirers. 30 cents each.

Philadelphus Coronarius. The Mock Orange; a medium sized shrub, bearing an abundance of white and sweet scented flowers; last of May. 25 cents each.

Spirea. Reevesii Flora Plena, Charming shrub with narrow, pointed leaves; blooms in May. 25 cents each.

Tritoma Grandiflora. Few flowers produce such a striking effect and are so attractive as this grand herbaceous plant, which throws up stalks three to five feet high, bearing large and solid spikes of flame colored flowers of great size and brilliancy; grown singly or in masses it has a grand effect, blooming until snow comes, regardless of Fall frosts. 25 cents each.

Tamarix Gallica. The pink flowers of the Tamarisk, borne all along its slender branches, and its delicate feathery foliage, give it a character no other shrub possesses. 50 cents each.

Viburnum Plicatum. The Japanese Snowball; a beautiful shrub of moderate upright growth, with crinkled or plicated rich green leaves; the flowers are white, and larger and more solid than those of the common Snowball. 25 and 50 cents each.

Wiegelias. Beautiful shrubs that come in flower in June and July; the flowers are produced in so great profusion as to almost entirely hide the foliage; they are very desirable for the border or for grouping, and also as specimen plants for the lawn; the flowers are rose and white, and of very pretty form. 25 cents each

Our Surprise Collection.

To meet the demand of our large trade we grow a very large stock of plants, in most cases more than needed, to make sure of having enough. After business is about over, in May, we make this surplus up into "Surprise" Collections, and offer them to our customers, giving them three or four times the value of their money. The "Surprise" comes in the great amount of fine plants you get for the money paid. These collections cannot be sent out until about May 10th, and will be entirely our own selection from kinds of which we have a surplus. The purchaser can state the purpose for which they are wanted, and we will select accordingly. No collection will be made up for less than ONE DOLLAR, and from that up to any amount the purchaser may desire. In all cases one will be surprised at the amount of fine plants he gets for the money. All "Surprise" Collections ordered previous to May 10th will be booked and shipped then, but we can continue sending them through May and June. Our "Surprise" Collections will give unbounded satisfaction. They are indeed a surprise. Sent only by express.

ORNAMENTAL CLIMBERS.

TWO NEW CLIMBERS.

The Two for $1.00.

SCHUBERTIA GRANDIFLORA.

A new plant of great merit, destined to take a prominent position as one of the best novelties introduced for many years. It flowers profusely in clusters, shape of the Allamanda, pure white, deliciously scented, and lasting a long time in water; flowers large and of good substance. For cut blooms and decorative purposes it is unique; culture most simple; hardy South of Tennessee. 75 cents each.

CENTROSEMA GRANDIFLORA.

A new hardy perennial climber growing six to eight feet the first season. The foliage dies to the ground in Winter and comes up with renewed vigor the following Spring. Flowers large, inverted pea-shaped; colors vary from white to purple with a white eye. Centrosema luxuriates in a poor sandy soil, so that almost any waste spot will suit it. 35 cents each.

APIOS TUBEROSA.

The Tuberous Rooted Wistaria. Valuable, hardy, tuberous-rooted climber, closely resembling the common Wistaria in vine and foliage, and having clusters of rich, deep purple flowers, which have a strong, delicious Violet fragrance. 15 cents each.

ALLAMANDAS.

The Allamandas are beautiful evergreen climbers, with rich, glossy foliage, and deep yellow flowers, which are very large and showy; it would be difficult to exaggerate the beauty of the Allamanda or its real and permanent value. 25 to 50 cents each.

AKEBIA QUINTATA.

A climbing plant from Japan, with a beautiful cut foliage, having large clusters of chocolate colored flowers, which are very fragrant; attains the height of twenty feet. 25 cents each.

AMPELOPSIS QUINQUEFOLIA.

Rapid grower, attaching itself to brick, stone walls or to trees; it has beautiful green foliage in Summer, turning to a rich crimson in the Autumn. 25 cents each.

AMPELOPSIS VEITCHII.

Miniature variety of Virginia Creeper, which clings to any building and produces dense foliage of glossy pale green, shaded with purple, and which turns a brilliant red in the Autumn. 25 cents each.

ARISTOLOCHIA SIPHO.

Dutchman's Pipe. Very large leaves and brownish flowers of a very singular shape, resembling a pipe; it is a vigorous and rapid growing climber, attaining a height of twenty feet.

CLERODENDRON BALFOURI.

Very handsome greenhouse climber, with large clusters of crimson scarlet flowers, each flower encased in a bag-like calyx of pure white. 25 cents each.

CORÆA SCANDENS.

Magnificent climber, with large, bell-shaped flowers and elegant leaves and tendrils; it is of very rapid growth, and consequently eminently adapted during the Summer for warm situations. 20 cents each.

CISSUS DISCOLOR.

A well known climber, with the leaves beautifully shaded dark green, and white, the upper surface of the leaf having a rich velvet-like appearance. 15 to 30 cents each.

HOYA CARNOSA.

Wax Plant. Has thick, fleshy leaves, growing moderately fast and bearing umbels of beautiful flesh-colored flowers and from which are exuded large drops of honey-like liquid; one of the best plants for house culture, as it stands the extremes of heat and cold better than most plants, and is not easily injured by neglect. 25 cents each.

IPOMŒA PANICULATA.

Mexican Morning Glory. One of the most attractive climbers; flowers as large and plentiful as those of the Moon-flower; plant a rapid, vigorous grower; tuberous rooted; can be lifted in the Fall and stored like a Dahlia; color a beautiful shade of rosy lilac; planted with the Moonflower, they unite to make a grand display; the pure white flowers of the one appearing in the evening, and those of both remaining fully expanded for several hours in the morning. Strong roots. 25 cents each.

IPOMŒA NOCTIFLORA.

The Evening Glory, or Moonflower. There are few plants we have seen sent out that have been so satisfactory as this; one lady says that it was trained on strings to a balcony twenty-five feet high and forty feet wide, and that from August to November it was covered nightly with its white moon-like flowers from five to six inches in diameter; it has also a very rich Jessamine like odor at night. 15 cents each.

MANETTIA VINE.

A charming and profuse flowering climber, producing tubular flowers from one to two inches in length, the plant being literally covered with them the entire season; the coloring of the flowers is gorgeous in the extreme, being a flame color tipped with a bright yellow; the contrast with the vivid green, glossy foliage is startling, making the plant an object of rare beauty. 10 cents each.

PASSIFLORA QUADRANGULARIS

A magnificent variety; the flowers are very large and sweet scented; purple inside of petals, light green on the outside; the centre of the flower is of many colors. 25 cents each.

WISTERIA SINENSIS.

One of the most hardy climbing plants, and when once established, of very rapid growth, covering the entire side of the house in a few years, presenting a magnificent appearance when in full bloom. 50 cents each.

Select Hardy Evergreen Trees.

Arbor Vitæ, Tom Thumb. Very small, compact little Evergreen; a beautiful ornament for a small yard or cemetery lot. 50 to 75 cents each.

Arbor Vitæ, American. This plant is all things considered, the finest Evergreen. 25 to 50 cents each.

Arbor Vitæ, Pyramidalis. Exceedingly beautiful, bright variety, resembling the Irish Juniper in form. 50 to 75 cents each.

Hemlock Spruce. Remarkably graceful, beautiful tree, with fine drooping branches and the delicate dark foliage of the Yew. 50 cents each.

Irish Junipers. Erect and formal in its habit; is much used in cemeteries; 18 inches high, 25 cents each; extra large, $1.00.

Balsam Fir. Handsome tree of compact growth, globular in form; most desirable. 50 cents each.

Mahonia Aquifolia. Evergreens with bright shiny leaves and showy bunches of yellow flowers in the early Spring. 25 cents each.

Norway Spruce. Lofty elegant tree of perfect pyramidal habit; very popular; should be largely planted; one of the best Evergreens. 50 cents each.

Retinospora Plumosa. Exceedingly handsome Japanese Evergreen, with feathery, light green foliage. 25 to 50 cents each.

Retinospora Plumosa Aurea. Like the preceeding, a plant of great beauty, soft, plume like foliage. 25 to 50 cents each.

Our stock of Irish Junipers and Retinosporas are very fine and of good size for the low price asked. The Irish Junipers are of stately growth and always in demand. The Retinosporas are the most graceful of all the Evergreens in cultivation and of free and constant growth.

THE STRAWBERRY.

Of all the fruits which a bounteous Nature has provided for the use of man, none, we think, contains so many charms as the strawberry, none so nearly fills the requirements of a universal favorite. Coming at a time of the year when the human appetite is in its most capricious state, when the long months of absence of nearly all fresh fruit has created a peculiar craving for it, the strawberry fills a want which but for it would of necessity remain unsatisfied. Perhaps this want makes it more highly prized than otherwise it would be. Still, we cannot believe it would take a lower place if it came at any other season of the year. It is to fruits what the rose is to flowers; not so conspicious as some, not so hard to grow as some, but still the queen. So beautiful in form and color as to be an ornament to any table, so fragrant as to please the most fastidious nostril, in flavor so delicious and in healthfulness so good as to appeal to the most capricious taste or delicate stomach, it comes as a welcome visitor to the housewife, epicure and invalid, and grieves no one unless it be the doctor.

The varieties we offer this year are all first-class. A judicious selection from our list will give fruit from the earliest period of the strawberry season until late in the Summer. Crescent and Henderson are good early varieties, Bubach and Haverland have no equal for mid-season, and Gandy is the best late. We will make a selection from the following lot at $5.00 per 1000 to parties intending to plant largely, or will make special price on large lots upon application.

CRESCENT.

This is beyond question a wonderful berry. Has yielded as high as 8,000 quarts per acre. In size it is medium to large; in color brilliant and handsome. The fruit colors on all sides at once, so that all red berries may be gathered, a quality appreciated by all growers. It bears immense crops, even in weeds and grass. It is an "iron-clad" for the sun or rain, cold or heat. The great trouble with most growers is that they allow the Crescent to spread too much. In order to produce best results the rows should not be allowed to get over twelve to fourteen inches wide, and then if the plants get too thick in the rows, take a hoe and cut out some of the plants, allowing each plant three inches of ground. By so doing you will get better fruit and more quarts to the acre. 25 cents per dozen; 75 cents per hundred.

LOVETT'S EARLY.

The introducer describes this berry as follows: The fruit is of the largest size, long and little shouldered, bright scarlet, with smooth surface and yellow seeds; in quality it is as good as Sharpless and stands shipping well; it colors all over at the same time, as early as the earliest, with a perfect blossom; and for vigor, hardiness and productiveness it surpasses other varieties; originated in Kentucky. The originator claims it is as heavy a yielder as the Crescent, and the berries twice as large in size. 25 cents per dozen; 75 cents per hundred.

GANDY.

This is the best and finest late Strawberry yet introduced, and we find it to be the favorite late variety with fruit growers all over the country. By its use the season of Strawberries is extended by nearly two weeks. The berries are of large, uniform size and shape, of bright crimson color, are very handsome and showy, of superior quality, very firm, and ripen two weeks after Sharpless. For the home table it is invaluable, its fine quality, beauty and size rendering it a universal favorite. 25 cents per dozen; $1.00 per hundred.

PARKER EARLE.

This magnificent new berry originated in Texas and is named in honor of Mr. Parker Earle, the distinguished President of the American Horticultural Society. It produces wonderful crops on light soils and endures hot, dry weather better than most Strawberries. The plant is a robust grower with perfect flowers, berries large, conical, regular and uniform, glossy crimson, flesh firm, reddish and in quality excellent. It is remarkably vigorous on all soils and under all conditions; early to medium. 25 cents per dozen; $1.00 per hundred.

LADY RUSK.

The foliage is entirely free from rust or disease; color a bright and handsome scarlet, and of most excellent flavor; the fruit of the Lady Rusk has been shipped all over the northwestern states, as far as Winipeg, Manitoba, in perfect condition; the nature of the fruit after standing for ten days or two weeks is to dry up, having the appearance of evaporated Raspberries or Blackberries, very good points in its favor during wet seasons at picking time. It does not melt down or rot like other varieties, but dries up; the fruit is of large size, several days earlier than the Crescent, and is its superior in every respect, holding size well throughout the entire picking, and, above all, is one of the most productive varieties. 25 cents per dozen; $1.00 per hundred.

HAVERLAND.

A magnificent and comparatively new variety; very early and productive, particularly in rich, deep soil. It is exceedingly vigorous and healthy, bearing very large and handsome fruit of fine quality. The shape of the berries is rather long, and of a glossy and bright crimson. 25 cents per dozen; 75 cents per hundred.

BUBACH NO. 5.

A superior variety in every respect, of large size and great productiveness; the plant, too, possesses great vigor and is of strong growth. A prominent Strawberry grower says he has tested every variety sent out in the last twenty years, and if all varieties were culled out but ten, he would place Bubach's No. 5 at the head of the list on account of its large, bright, well-colored berries, its immense yield and handsome appearance and fine flavor. 25 cents per dozen; 75 cents per hundred.

MICHEL'S EARLY.

The originator says this is the only Strawberry entitled to be called "early," being two weeks earlier than any other variety. It is a robust, strong grower, rooting deeply, and throwing large and luxuriant foliage on strong stems, with never a trace of rust or mildew; it yields bountifully, being as productive as the Crescent; berries medium to large, of a handsome, regular form; color a bright scarlet; flavor exquisite, pronounced by all who have tasted it equal to that of the wild Strawberry. 25 cents per dozen; $1.00 per hundred.

SHARPLESS.

Fruit large to very large, an average specimen measuring one and one-half inches in diameter; in form it is generally oblong, narrowing to the apex, and irregular and flattened; color a clear and bright red, with a shining surface; flesh firm and sweet; plant very vigorous. 25 cents per dozen; 75 cents per hundred.

THE "HENDERSON."

The fruit is of the largest size, early, and immensely productive, but its excelling merit is its exquisite flavor; it is a strong grower and stands the dry Summers well. 25 cents per dozen; $1.00 per hundred.

Grapes, Raspberries, Etc.

GRAPES.

We grow an assortment of the most popular Grape vines. Fine two year old plants, 15 cents each; $1.50 per dozen.

RASPBERRIES.

50 cents per dozen; $2.00 per hundred.

Turner. Very hardy, which character makes it the favorite in the South.

Gregg. One of the best and largest.

Hansel. One of the earliest; very bright scarlet.

Cutberth. Rich and luscious; crimson;

Schaffer's Colossal. New overbearing Raspberry.

BLACKBERRIES.

Early Harvest. Very early.

Kittatiny. Large berry.

Snyder. Hardiest of all. 50 cents per dozen; $2.00 per hundred.

GOOSEBERRIES.

Downing. Very large, handsome green, of splendid quality for both cooking and table use. 20 cents each; $2.00 per dozen.

Houghton Seedling. Small to medium; pale red, roundish oval; sweet, tender, and very good. 20 cents each; $2.00 per dozen.

CURRANTS.

Large, two years, 15 cents each; $1.50 per dozen.

Red Dutch. Old reliable sort.

White Grape. The best white.

Black Naples. Good old variety.

Vegetable Plants.

ASPARAGUS.

The preparation of the Asparagus bed should be made with care, from the fact that it is a permanent crop which ought to yield well for twenty-five years. The ground must be thoroughly drained; light sandy loam is preferable. Work in about six inches of manure, two feet deep, as the roots of the plant will reach that depth in a few years. The crowns of the plants should be placed at least three and a half inches below the surface. The surface of the bed should have a top-dressing of three-fourths inches of rough stable manure every Fall. Salt is also a good manure. Plant in rows eighteen inches apart, and set the plants nine inches apart.

Conover's Colossal. A mammoth variety of vigorous growth, sending up from fifteen to twenty sprouts each year from one to two inches in diameter; color deep green, and crown very close. Two year old roots, $1.00 per hundred.

Palmetto. Of Southern origin; has now been planted in all parts of the country, and reports indicate that it is equally as well adapted for all sections. It is earlier, a better yielder, more even and regular in its growth, and in quality equal to that old favorite, Conover's Colossal. Two year old roots, $1.00 per hundred.

SWEET POTATO PLANTS.

Thirty cents per hundred; $2.50 per thousand.

CABBAGE PLANTS.

Nice young plants ready March 1st in all the leading vraieties at 25 cents per hundred; $2.50 per thousand.

TOMATO PLANTS.

These we grow in quantities in Nepon set paper pots, and can be shipped any distance without breaking the ball or injury to the plant. Pot grown Tomato plants are from two to three weeks earlier than tender plants from frames and hotbeds, that have to be dug up without a ball of dirt to the roots, and consequently wilt and suffer for several days after planting before they take a fresh root. On the other hand the pot grown plants never receive a check or set back in any way, but keep right along growing and fruit two weeks in advance of hotbed and frame grown plants. Those that have once tried POT GROWN TOMATO PLANTS would never use any other. Try a dozen and be convined. Ready February 1st. We grow all the leading varieties. By express, 50 cents per dozen; $4.00 per hundred.

RHUBARB, OR PIE PLANT.

$1.50 per dozen.

CELERY PLANTS.

Ready in June. 50 cents per hundred.

EGG PLANTS.

Twenty-five cents per dozen.

PEPPER PLANTS.

Ten cents per dozen.

CUT FLOWERS.

We have at all seasons a quantity of beautiful cut flowers and designs suitable for all occasions. If you want a funeral design, marriage offering, or commencement bouquet, or basket or box of loose cut flowers, write or telegraph us, stating the purpose for which it is required, and we will send something both pretty and appropriate.

BEAUTIFUL FOWLS.

BRONZE TURKEYS.

In addition to our large stock of beautiful Pea Fowls, we have now a flock of magnificent Bronze Turkeys of the finest breed, and offer a limited number of fowls and eggs this Spring. Where these fine birds have plenty of grass, they do not disturb the vegetable garden, but destroy immense numbers of cabbage and tobacco worms, and all kinds of insects. Prices of our Bronze Turkeys, cooped and delivered at express office: For a pair of large fine fowls, $9.00; for a trio of large fine fowls, $12.00; for 13 eggs, packed, $5.00; for 7 eggs, packed, $3.00.

PEA FOWLS.

We have a limited number of these highly ornamental birds that we take great pleasure in offering to our patrons. There is nothing more beautiful on large grounds than the brilliant plumage of Pea Fowls strutting among the evergreens and trees. Everybody that visits our place are delighted with the effect. Our stock has grown too numerous for our place at present, and we wish to dispose of some them. The birds have to be three years old before their long and beautiful tail is fully grown and their plumage at its prettiest. They live for many years; we have some here fifteen years old, and their plumage is as gay as ever. Price, per pair, one year old, $5.00; two years old, $10.00. Peacocks, three years old, with full tail and handsome plumage, $7.00. Can be shipped in crates by express.

LAWN GRASS MIXTURE.

Our preparation of Lawn Grass has given great satisfaction to all who have used it. It produces a beautiful lawn and stands the Summers admirably. Finest prepared, per bushel, $3.00; per peck, $1.00.

FLOWER POTS.

These are carefully packed in barrels and delivered to the express or freight office, and receipted for, they are then at purchaser's risk.

2½ inch, 40 cents per dozen, $2.50 per 100. 8 inch, 15 cents each, $1.50 per dozen.
4 inch 60 cents per dozen, $4.00 per 100. 10 inch, 25 cents each, $2.50 per dozen.
6 inch, $1.00 per dozen, $7.00 per 100. 12 inch, 50 cents each, $5.00 per dozen.
7 inch, $1.25 per dozen, $8.00 per 100.

GOLD FISH.

The Gold Fish was introduced into the United States many years ago from China. In recent years a superior stock has been introduced from Japan, possessing more brilliant color and of fantastic shapes. Some are red and white (pearl); others of all or part of these colors and mottled with black. In shape, some are long and some are round bodied, some have fan-like tails; some straight tails, and others three-lobed tails, etc. There is nothing more attractive in a room than Gold Fish, and there is but little trouble in keeping them. Change the water two or three times a week, river or cistern water will do, and clean the globe once a week. Feed the fish wafer crackers, a little at a time. They can be sent any distance by express in tin cans at buyer's risk. We have collected a very interesting stock of these favorite little household pets. We will send a handsome pair of Gold Fish and a large two-gallon glass globe, amply large enough for a pair of our largest fish, for $3.00, securely packed, by express.

PLAIN GOLD FISH.

Bronze, Red, Pearl and Variegated. Small young fish, 25 cents each; finely-colored plain Gold Fish, 50 cents each.

CHINESE MOON FISH.

These are the fringed tailed Gold Fish. Superior to the ordinary Gold Fish by their large fringed tail of most gorgeous colors. The most graceful of all fish. They are rich bronze, red, pearl, black-tipped fins and variegated. 75 cents each.

CHINESE FAN TAILS.

These are brilliant and almost transparent fish with a most graceful tail, triangular in form, having three lobes or, as it appears, three different tails. The most graceful of all fish. $1.00 each.

THE PARADISE FISH.

This Fish has only recently been introduced into the United States. As the Bird of Paradise is the most graceful and beautiful among the feathered tribe, so this may well be considered the most beautiful among the fishes. It is a native of Cochin China, or rather it comes from that country where it is cultivated solely for ornamental purposes. It is said to be found nowhere in a wild state. This fish is peculiarly well adapted for aquariums requiring very little water. The male, which is most brilliant in color, does not exceed, when full grown, four inches in length. The predominant color is bluish-green, the other colors are orange, red, gray and black. The large fins, (dorsal and ventral) also the tail, are very beautifully marked, especially when the sun is shining on it it seems to give out every imaginable color. The tail when expanded is very large in proportion to the size of the body. The female is smaller than the male, and duller in color. It is not only the brilliant coloring of the fish which makes them attractive as house pets. One cannot fail to be struck by their strange motions when compared with other fish, and having once observed their methods of keeping house as it were, we cannot help admiring their wonderful instinct. The male is the active or working partner. When the female is about to lay her eggs this is indicated by the great activity of the male. He sets to work and builds a nest in a secluded part of the aquarium, generally under some floating leaves or some floating plants. The nest is composed of an immense number of air bubbles, which is built in this way. The male comes to the surface, takes a mouthful of air, goes about three inches below the roof of the nest and then ejects the air in small bubbles which naturally rise until they come in contact with the leaves or floating plants. He repeats this process until the nest is about one-fourth of an inch thick, when it is completed he guides the female to the vicinity of the nest where she deposits the eggs, the male then gathers the eggs in his mouth and places them among the air bubbles. This finished, he chases the female away, for the reason she devours the eggs whenever she gets the chance. The male now seems to be well aware of the responsibility resting on him, guarding them from enemies and supplying them with food until they are old enough to take care of themselves. Price $3 per pair.

FISH GLOBES.

We have a nice lot of Globes suitable for fish of all kinds, large enough to hold two gallons of water. Standing Globes on stout pedestal that make a handsome ornament in a hall or sitting-room, $2.00 each, packed securely for express.

Shipping cans for fish, 15 cents each for two to six fish, larger cans in proportion. As many as one hundred are shipped in a ten gallon can, with about eight gallons of water. Cans charged at cost price, or money for same will be refunded if they are returned with expressage paid.

CHRYSANTHEMUM POTS.

We have a great demand for large pots for growing specimen Chrysanthemums in, and as these large pots are heavy, easily broken, and expensive, we here state for our customers' benefit that an ordinary water bucket, with the handle taken off and a few holes bored in the bottom, makes an ideal Chrysanthemum bucket. If painted nicely on the outside they are even more ornamental than an ordinary flower pot, as well as being light, cheap and not easily broken. We grow hundreds in such buckets, and everybody admires them, and the idea is fast being taken hold of by all Chrysanthemum growers. We have the buckets all ready for use. Being light, unlike pots, they can be sent by express. Price $1.50 per dozen.

Chrysanthemum Culture in America!

BY JAMES MORTON.

An excellent and thorough book; especially adapted to the culture of Chrysanthemums in America. The contents include

Oriental and European History.	Disbudding and Thinning.
American History.	Sports and Variations.
Propagation by Cuttings.	Standard Chrysanthemums.
Propagation by Grafting, Inarching and	Insects and Diseases.
Seed.	Early and Late Flowering Varieties.
General Culture. Exhibition Plants.	Chrysanthemum Shows, etc.
Classification. Exhibition Blooms.	List of Synonyms.
Soil for Potting. Selection of Plants.	Varieties for Various Purposes.
Top Dressing. Staking and Tying.	Hints on Exhibitions.
Chrysanthemums as House Plants.	National Chrysanthemum Society.
Watering and Liquid Manure.	Monthly Calendar.

"The book is full of interest to all who admire what the author calls The Star-Eyed Daughter of the fall, and contains enough valuable information for florists to insure for it a place in their library. The book will certainly be welcomed by all florists."—Florists' Exchange.

"Much valuable information is brought together in this book on all matters pertaining to the Chrysanthemum, and must prove very acceptable to American growers of the Autumn Queen."—American Florist.

Illustrated. Price, Cloth, $1.00; Paper, 60 Cents.

If both our books are ordered together, we will make the price for the two.

CHRYSANTHEMUM CULTURE for AMERICA and SOUTHERN FLORICULTURE

ONLY ONE DOLLAR AND FIFTY CENTS.

Order at once, and books will be mailed free on receipt of price.

Address,
JAMES MORTON,
CLARKSVILLE, TENN.

www.ingramcontent.com/pod-product-compliance
Lightning Source LLC
Chambersburg PA
CBHW022010190326
41519CB00010B/1470